The Rise of Tourism in China

TOURISM AND CULTURAL CHANGE

Series Editors: Professor Mike Robinson, *Nottingham Trent University, UK* and Professor Alison Phipps, *University of Glasgow, Scotland, UK*
Associate Editor: Dr Hongliang Yan, *Oxford Brookes University, UK*

Understanding tourism's relationships with culture(s) and vice versa, is of ever-increasing significance in a globalising world. TCC is a series of books that critically examine the complex and ever-changing relationship between tourism and culture(s). The series focuses on the ways that places, peoples, pasts and ways of life are increasingly shaped/transformed/created/packaged for touristic purposes. The series examines the ways tourism utilises/makes and re-makes cultural capital in its various guises (visual and performing arts, crafts, festivals, built heritage, cuisine, etc.) and the multifarious political, economic, social and ethical issues that are raised as a consequence. Theoretical explorations, research-informed analyses and detailed historical reviews from a variety of disciplinary perspectives are invited to consider such relationships.

All books in this series are externally peer-reviewed.

Full details of all the books in this series and of all our other publications can be found on http://www.channelviewpublications.com, or by writing to Channel View Publications, St Nicholas House, 31-34 High Street, Bristol, BS1 2AW, UK.

TOURISM AND CULTURAL CHANGE: 62

The Rise of Tourism in China

Social and Cultural Change

Yiping Li

CHANNEL VIEW PUBLICATIONS
Bristol • Jackson

DOI https://doi.org/10.21832/LI8908
Library of Congress Cataloging in Publication Data
A catalog record for this book is available from the Library of Congress.
Names: Li, Yiping, author.
Title: The Rise of Tourism in China: Social and Cultural Change/Yiping Li.
Description: Bristol, UK; Jackson, TN: Channel View Publications, 2023. | Series: Tourism and Cultural Change: 62 | Includes bibliographical references and index. | Summary: "This book offers a comprehensive understanding of China's tourism development from 1992 onwards, focusing on the social-cultural change that accompanied the rise of tourism. It examines both the economic benefits and sociocultural impacts of tourism and argues that a delicate balance between these is needed to achieve sustainable tourism"—Provided by publisher.
Identifiers: LCCN 2022053549 (print) | LCCN 2022053550 (ebook) | ISBN 9781845418908 (hardback) | ISBN 9781845418922 (epub) | ISBN 9781845418915 (pdf)
Subjects: LCSH: Tourism—China. | Tourism—Social aspects—China. | Tourism—Economic aspects—China.
Classification: LCC G155.C55 L5276 2023 (print) | LCC G155.C55 (ebook) | DDC 306.4/8190951—dc23/eng20230323
LC record available at https://lccn.loc.gov/2022053549
LC ebook record available at https://lccn.loc.gov/2022053550

British Library Cataloguing in Publication Data
A catalogue entry for this book is available from the British Library.

ISBN-13: 978-1-84541-890-8 (hbk)

Channel View Publications
UK: St Nicholas House, 31-34 High Street, Bristol, BS1 2AW, UK.
USA: Ingram, Jackson, TN, USA.

Website: www.channelviewpublications.com
Twitter: Channel_View
Facebook: https://www.facebook.com/channelviewpublications
Blog: www.channelviewpublications.wordpress.com

Typeset by Deanta Global Publishing Services, Chennai, India

Contents

Acknowledgements

This book would not have come to fruition without the supports for my research projects of the past 30 years, which were offered by the University of Alberta, the University of Western Ontario and Hong Kong Hui Oi Chow Trust Fund.

I would like to thank my research students Chammy Lau, Connie Kwong, Kun Lai, Zhiyi Hu and Zixin Mai, who were excellent assistants to my research projects through all those years when I worked at the University of Hong Kong and Guangzhou University. I hope they know how appreciated they are.

I would also like to thank my supervisor at the University of Alberta, Canada – Professor Tom Hinch, a gentleman who opened for me a fascinating world of tourism research 30 years ago.

In particular, I would like to thank my remarkable parents whose nurturing made me an authentic human being.

Figures and Tables

Figures

Tables

Preface

This book is an attempt to integrate discourse and place theory in studying the tourism phenomenon in China. By doing so, I intend to join the efforts of other scholars to challenge and dismantle the scientific-positivist paradigm that is prevalent in tourism research. My ultimate goal is, especially, to explore the possibility of pushing tourism research within China from a positivist spatial science that seeks objective facts, to a more humanistic field of study involving existential and phenomenological methods. My motto of life has always been that the true wanderer, whose travels are happiness, goes out not to shun, but to seek (Stark, 1953). From childhood, I have been fascinated with faraway places and everything foreign – the ways that different cultural groups think, behave, create and express themselves. My empathic nature has led me, over the past 30 years, to venture into the career of a tourism researcher and a world traveller. As a geographer, I have striven to make the self appear strange and the other familiar. The emotional rewards of travel accrue as the associated inquiries go, but the humanitarian concerns always lie beneath the surface. A question recurs in the mind: can leisure travel lead human society to an altered state, or offer people some kind of aspiration for a renewal, to turn the world into a better place? This question continues, increasingly puzzling nowadays, as the COVID-19 pandemic seems endless in this nation and, in the meantime, Ukraine and Taiwan – two crises, 5000 miles apart – seem linked in complex ways.

Introduction: Making or Remaking People and Places through Tourism

Tourism has played a key role in China's social and cultural arena since 1978 when the central state began to allow the nation to embrace the world economy. After the recession inflicted by the 1989 Tiananmen Square incident, Visit China '92 – the first such event ever in the history of the country – ushered a milestone for the tourism industry. Tourism is thus mobilised as an ideological vehicle for implementing the nation-state's economic policies, whereas the people and places have undergone tremendous change in the process of the nationalistic discourse construction. This book is a collection of my 30-year research and reflection of tourism in China. Regarding tourism as a discourse of difference and based on empirical research, the chapters in this book aim to offer readers a comprehensive understanding of China's tourism industry over the past three decades, and the associated social and cultural change.

When the People's Republic of China was founded on 1 October 1949, the country was poor, reclusive and isolated from the outside world. Supplies of basic commodities were so tight that the consumption of tourism and leisure services was out of the question. This all changed when the late paramount leader Deng Xiaoping initiated economic reform and opening-up in 1978. Tourism afterwards was considered both a gesture of embracing the world and one of the driving forces for economic reform. The country's decades-long isolation, however, created both sociocultural and physical distances to the outside world. From this point of departure, this book chronologically reviews the rise of tourism in China and the associated social and cultural changes in the past three decades.

Chapter 1 explores how the appeal of those distances became an enticement to international tourists, thereby boosting China's inbound tourism in the last decade of the 20th century. It presents an empirical exploration of international tourists' travelling in China during that period of time. Through a detailed description of the lived experiences of tourists on two separate package tours – one in 1996 and the other in 1997 – I attempt to scrutinise whether and to what extent geographical consciousness influences international tourists' experiences of China.

The purpose is firstly, to uncover the appeal of distance that China drew international tourists to visit the country and secondly, to provide insights for research and application in this line of tourism study.

Chapter 2 directs readers' attention towards ethnic tourism in the mid-1990s. Two specific ethnic tourism sites – Yunnan Folk Cultural Village of China and Wanuskewin Heritage Park of Canada – are selected for case studies, in order to debate the issue of the influence of Said's 'Orientalism' on traditional ethnic tourism practice. Through a comparative approach, the debate is directed to the development processes, social/cultural issues and development strategies, which are manifested in the ethnic tourism practices of the two case sites. By utilising predominantly qualitative methods, including documentary research, in-depth interviews and participant observations, this chapter intends to provide insights into sustainable development strategies of this type of tourism practice during that particular period of time.

After two decades of development, tourism began to show various impacts – the good, the bad and the ugly – on Chinese society. Chapter 3 delineates my observation and reflection of those impacts on a Chinese local community when the new millennium began. The chapter delineates the evolution of China's tourism development through a comprehensive review of relevant tourism policy documents first, following up with empirical observation of some specific tourism activities. By doing so, a conceptual model is proposed, regarding tourism as a fashion industry that evolves with consumers' changing perceptions, expectations, attitudes and values. In this vein, I argue that maintaining a sustainable social-cultural, economic and natural environment is imperative for both the survival and progress of destination communities.

Chinese tourism scholars began to apply the concept of community tourism as early as the late 1990s, but they understood the concept rather differently from their Western counterparts. Chapter 4 highlights some contradictions that are manifested in China's modernisation process to discuss those differences. Using in-depth analyses of two specific cases of tourism development through 2002–2005, this chapter points out that the differences basically hinge on two questions: is community tourism a democratic process or a planning strategy? and is sustaining community or tourism the overall objective?

So-called red tourism became popular in China around the start of the 21st century. Its patrons have continued to grow over the past decades with the ever-increasing financial and administrative support of the nation-state. Chapter 5 presents a comprehensive picture of China's red tourism development in order to provide insights for quality research on heritage tourism in general and red tourism in particular. Based on two factual elements, I attempt to differentiate red tourism from the counterpart phenomenon in Eastern Europe – so-called communist heritage tourism. This differentiation is to facilitate an in-depth exploration of red tourism practice in China through 2006–2010, thereby addressing the political uniqueness of red tourism and its association with some emerging issues of heritage tourism research within China.

China's nation-state regards the 2008 Beijing Olympic Games as a symbolic momentum for the rise of the country. Nearly the entire nation was mobilised to host this mega-event after Beijing was selected as the host city on 13 July 2001. Chapter 6 presents a case study that explores the roles of this mega-event in affecting or changing China's capital city Beijing. The study results show that the Olympic Games helped transform Beijing as an upgraded tourist destination via the enhancement of the city's tourism elements and the associated spatial organisation, the external linkage with adjacent destinations and visitors' subjective appraisals of Beijing.

After one decade of reunification with mainland China, Hong Kong still stands largely different, social-culturally and politically, from the motherland due to its century-long colonial legacy. Chapter 7 touches upon some contested issues regarding the relationship between China and Hong Kong by scrutinising the experiences of both residents and China's mainland tourists in sharing the shopping space of Hong Kong. The chapter specifically aims to examine the relationship between Hong Kong residents and mainland Chinese tourists, in order to contribute to the knowledge pool of host–guest relationship research in tourism. This chapter concludes that a dual identity could be attached to an individual simultaneously in one place in the host–guest interaction, while the politics of mobility exists within the power struggle that influences the in-place/out-of-place status.

Chapter 8 explores the essence of festival meanings – the extent to which the perceived sociocultural benefits of festivals articulate the production of a sense of place (SOP) and its respective locality. By comparative analyses of two specific community festivals in Hong Kong through 2013–2016, it presents local residents' views of the festivals. The chapter suggests that festival meanings be regarded as crucial elements that are attributed to SOP, which eventually shape the identification of place uniqueness. Tourism representations, therefore, are not only passive reflections of prevailing cultural values drawn from current stereotypes and images, but they also play a prominent role in shaping values, behaviours and identities by contributing to the socialisation process.

Chapter 9 focuses on Chinese domestic tourists' holiday experiences on a resort island off the southern coast. It presents a specific case study that was conducted through summer and autumn of 2020, in order to explore the dynamic relationship between the linguistic landscape and the island's place making. Using in-depth examination of the language on public transport signs, billboards, street names, shop names and commercial posters, as well as public signs of government agencies of the island, the chapter interrogates the extent and characteristics through which place meaning is 'translated' into the linguistic landscape of a tourist island.

Chapter 10 explores the personal and collective urban nostalgia that shapes the performativity of domestic tourists, by analysing how tourism provides a space through which urban tourists understand the countryside, against all the odds of change in contemporary Chinese society. The chapter examines the social-cultural arena through which tourists experience rural China, thereby delineating how toured places are imagined and consumed. Through the lens of tourism, this chapter leads readers to ponder the social-cultural change post China's economic reform that began in 1978. It argues that the past is remembered in terms of the grand narratives of a bygone era of sociopolitical coherence, which tourists feel has been eroded by China's changing urban landscape.

The conclusion to this book highlights the theoretical and methodological lines that connect the chapters, namely the application of ethnography in geographical analyses of tourism in China. There is no single procedure to follow when using ethnography due to its recontextualisation over time through diversification across disciplines. The conclusion highlights approaches to the ethnographic enquiry of tourism that are predominantly adopted through the chapters of the book. Its ultimate goal is to join the efforts of other scholars to challenge and dismantle the scientific-positivist paradigm that is prevalent in tourism studies, and attempt to push tourism research within China from a positivist spatial science that seeks objective facts, to a more humanistic field of study involving existential and phenomenological methods.

All in all, this book offers readers a timely intervention into China's tourism development and the associated social-cultural change over the past 30 years. The 11 chapters are uniquely constructed to differentiate the book from other similar titles that have been published in the recent past. In a nutshell, the book is different and better because, firstly, it delineates, systematically and comprehensively, the tourism phenomena in China and the associated sociocultural change during the most recent decades (1992–present). Secondly, it is the first book of China tourism studies that takes a predominantly qualitative approach to the analysis of the phenomenon in question, as well as the first such book that covers the China–Hong Kong relationship in depth. Thirdly, the book fits well with the Channel View Publications' series – Tourism and Cultural Change. Last but not least, tourism is after all a fashion industry, whereas China is changing at an unprecedented speed. Timely knowledge and information on the change are especially needed, with which the author feels obliged to provide.

In terms of the intended readership, the book can be collected by university libraries as teaching materials and reference books for learning and research of relevant majors. In particular, its aspects about the people and places of contemporary China have high reference value for scholars and students in the field of human geography. Its focus on the social-cultural issues of tourism in China will particularly interest teachers and students of tourism management, leisure and recreation, and social sciences, for teaching and learning specific courses such as regional cultural geography of China, social and cultural issues in China's tourism industry, China's tourism resources and management, qualitative methods in tourism and leisure research, as well as modernisation, tourism and China's international connections.

1 The Appeal of Distant Places: China's Inbound Tourism in the 1990s

Introduction

The People's Republic of China was founded on 1 October 1949. It was a poor and reclusive country, isolated from the outside world. Basic commodities were in such short supply that tourism and leisure services were out of the question. This all changed when the late paramount leader Deng Xiaoping initiated China's reform and opening up in 1978. Tourism was then considered both a means of embracing the world and a vehicle for advancing economic reform (Zhao & Liu, 2020). However, China's decades-long isolation after 1949 created distinctive social-cultural and physical distances from the rest of the world. Wang (2000) points out that tourism is founded on the appeal of distant places that entice tourists to travel across spatial expanses to seek out different experiences. This point of view is supported by other Chinese tourism scholars. Travel distance is found to be critical among the factors influencing tourists' behaviour and their choice of destinations (Cao *et al.*, 2019; Xue & Zhang, 2020). The appeal of distance, and the social, economic, spatial and imagined facets that are central to creating real and imagined differences, are keys to understanding the ways in which international tourists desired to visit China during the last two decades of the 20th century.

This chapter presents a phenomenological exploration of the leisure tourism experience of international tourists travelling to and within China on two separate package tours – one in 1996 and the other in 1997. The purpose of the research is to provide an understanding of the influence of geographical consciousness on tourists' perceptions of China during the last decades of the 20th century. Place theory is utilised to analyse the performativity of tourists, in order to describe, in detail, their perceptions of the people and places of China during that time period. By doing so, the author attempts to uncover how travel experience connects China and the world through the processes of tourist imagination and consumption and how contemporary discourses of difference fuse with individual histories to formulate the ways in which China is understood by international tourists during that particular time period.

Literature Review

There are rich studies on the experiential features of geography, such as those on the experiences of place, space and landscape – both pleasant and unpleasant that people have, regardless of whether they know geography as a formal science (Billinge, 1977; Seamon, 1979, 2000; Taşer, 2022; Tuan, 1989, 1993, 2006, 2013). These experiences are the substance of human's involvement in the world, and constitute the phenomenological basis of geographical consciousness (Koelsch, 2004; Relph, 1990). In other words, geographical science presupposes a world that can be understood geographically and also that humans can feel and know themselves to be tied to the earth (Cifçi & Köybaşı, 2017; Desbois, 2016). Geographical consciousness has been recognised by many. Relph (1976, 1990) seeks geographical patterns regarding the immediate experience of life. Lowenthal (1961) argues that everyone is a geographer in a world practically geographical. This world is a 'life world': the sum total of a person's first-hand involvement with the world in which the person typically resides (Buttimer, 1976). This 'life world' theme also embraces the core of Tuan's (1977, 1989, 1993, 2001a, 2001b, 2013) studies on experiences of space, place, surface phenomena and aesthetic experiences of nature and culture.

The above highlighted experiential features of geography inform the main purpose of this chapter with three interwoven theoretical concepts – tourism experience, geographical consciousness and phenomenology. Opposing definitions of travel experience exist in tourism studies. Boorstin (1964) defines it to be a popular act of consumption and a contrived, prefabricated experience of mass tourism. In contrast, MacCannell (1973) believes it to be an active response to the difficulties of life in modern society, thus arguing that tourists are in search of 'authentic' experiences in order to overcome the difficulties prevalent in modern society. Both scholars, however, attempt to define the tourism experience with the notion that it has certain significance for modern individuals and for their societies. Those definitions initiate a debate over the negative and positive effects of modern leisure tourism (Furnham, 1984; Graburn, 1989; Sutton, 1967). The debate subscribes a common experience for all tourists as if their needs were consistent, regardless of the different social and cultural backgrounds constituting those needs. It is Cohen (1979) who launches the argument that different people need different experiences, casting different meanings for tourists and their societies.

Cohen defines the tourism experience as the relationship between a person and a variety of 'centres'. He believes that the meanings of the experience are derived from a person's worldview, depending on whether the person adheres to a 'centre'. The latter is not necessarily geographically central to the life-space of the person. It is 'the individual's spiritual centre which, for the individual, symbolises ultimate meaning'

(Cohen, 1979: 181). Cohen thus thinks that this experience reflects some stable patterns of motivation that both differentiate and characterise various modes of tourists' activities. Those activities are linked to the 'privately' constructed worlds of tourists and represent patterned ways of satisfying a wide range of personal needs – from pleasure to the search for meaning. Following this line of research, scholars allude to Cohen's modes of tourism experience in their studies (Hamilton-Smith, 1987; Nash, 1996; Page, 1997; Pearce, 1982; Ryan, 1991, 1997; Smith, 1989; Urry, 1990; Urry & Larsen, 2011; Yiannakis & Gibson, 1992). A general definition may be drawn from this commonality that the tourism experience is a multifunctional leisure activity involving either entertainment or learning or both, for an individual (Ryan, 1997; Sharpley, 2018).

In general, the learning involved is a natural process that mostly occurs incidentally (Kalinowski, 1992). It is experiential learning in an individual's spatial interaction with the tourist destination (Rahmat *et al.*, 2020), occurring in the individual's geographical consciousness. This geographical consciousness, in general terms, is the experience of place, space and landscape – both pleasant and unpleasant – that people have, regardless of whether they know anything about geography as a formal science (Billinge, 1977; Buttimer, 1976; Seamon, 1979; Tuan, 1989, 1993, 2013). This is the substance of an individual's involvement in the world and encompasses the emotion, the mind and the total self of the individual (Relph, 1990; Smith, 1982; Ven Passen, 1957). Geographical consciousness arises from the spatial and temporal bond between people and places. Sauer (1925) sees the land through the eyes of its inhabitants. Wright (1947) seeks 'geosophy' – geographical knowledge based on all points of view – in the place of imagination in geography. Lowenthal (1961: 260) views humans as 'artist and landscape architect, creating order and organising space, time, and causality in accordance with our perceptions and predilections'. Studies of the person–place bond by those scholars weave a discernible thread in the place theory of geographical science. In a nutshell, such reality is not at first an 'object', and geographical space is not a blank space waiting to be coloured and filled in. Some inquiries of objects, such as the sense of place and landscape, are not formalisations of a disciplinary perspective but root experiences that derive from a prescientific 'geographical consciousness' (Relph, 1990; Tuan, 1977, 1993). So, this knowledge should also be linked to forms of human experience that touch it in its immediacy, such as art, literature and history, working with and through humanities (Tuan, 1989, 2001a, 2001b). This requires geographers to possess the skills of expression, representation and careful description – the method of phenomenology (Buttimer, 1976; Cresswell, 2007; Mercer & Powell, 1972; Pickles, 1986; Relph, 1990; Smith, 1979).

The term 'phenomenology' comes from the Greek words *phainomenon* (an appearance) and *logos* (reason or word). Phenomenology is reasoned inquiry that discovers the inherent essence of appearance (Stewart, 1990). An appearance is anything of which one is conscious, and phenomenology is concerned with the question of consciousness. Anything that appears to consciousness is a legitimate area of phenomenological investigation. Phenomenologists reject natural sciences' quantitative methods of treating consciousness as an object, because it is not an object among others in nature and because there are conscious phenomena that cannot be dealt with adequately using the quantitative methods of experimental science (Spiegelberg, 1965). Phenomenological inquiry starts with the content of consciousness (whatever that content may be) as valid data for investigation (Stewart, 1990). It does not concentrate exclusively on either the object or the subject of experience, but on the point where being and consciousness meet.

Thus, the phenomenological method has undeniable advantages in studying the tourism experience. First, it is a human topic and its study requires a method such as phenomenology that does not limit its investigations to materialistic realities. Second, because experience is a 'felt perception', each individual must impart personal observations of the phenomena being experienced. When receiving this information, an investigator needs to exhibit flexibility, as they may confront unknown data (Thevenaz, 1962) on various perspectives of tourism experiences. The introspective approach of the phenomenological method facilitates the investigator to identify underlying dimensions. However, the phenomenological method has been largely overlooked in tourism research (Dann & Cohen, 1996; Li, 2000). The few studies that have probed the phenomenological side of tourism are theoretical and speculative (Cohen, 1979; Mannell & Iso-Ahola, 1987) or have only focused on the periphery of the subject, such as research on wilderness experiences (Morrison, 1986; Potter, 1993). So, utilising a phenomenological method to study the tourism experience can partially fill the gap in this line of research.

Research Methods

Phenomenological research focuses on human experience. It requires that the topics of interest are approached through the presence of conscious awareness (Manen, 2014; Polkinghorne, 1989). Instead of viewing the body as an organic object, phenomenological research examines the experience that people have of their bodies. In experience, events appear as meaningful, both the appearance of worldly objects or happenings and personal thoughts or feelings. Experience is meaningfully ordered, whereas its structure and order are difficult to describe. So, the essential task of a phenomenological method is to produce a clear, precise and systematic description of the meaning that constitutes the activity of consciousness in the human experience (Giorgi, 2009; Polkinghorne, 1989). In other words, the essence of the phenomenological method is to describe the meaning of an experience from the worldview of those who have that experience, and as a result attach a meaning to it (Kvale, 1996; Ray, 1994; Stewart, 1990). The method aims at a deeper understanding of a subject by descriptively illuminating its underlying structures. The goal of research is to describe rather than to explain the phenomenon in question. However, descriptive words may be imprecise, ambiguous or open to multiple interpretations. To carry out sound phenomenological research, therefore, researchers must be explicit about their data gathering and analysis procedures (Horrigan-Kelly *et al.*, 2016; Howe & Jackson, 1985; Riley, 1993).

Data collection

The majority of the data for this research were based on 39 international tourists' experiences on two separate package tours to and within China. One was conducted in November 1996 – a 25-day cross China tour with 9 tourists and the other was conducted in March 1997 – a 7-day tour of Beijing (via Hong Kong) with a group of 30 tourists. Questionnaire surveys and unstructured long interviews were adopted to collect the data. Participant observation was also utilised to detect the tourists' experiences through the unfolding processes. Three different questionnaires were developed before, between and after the two trips. The first questionnaire, designed before the research trips, was distributed to all the tourists. It was semi-structured with an 'informed consent form' to help the respondents understand that their participation would remain anonymous. This informed consent form also requested potential long-interview participants to provide their demographic information. From the first trip, five tourists were selected for long interviews. The second trip was taken four months later, and a new questionnaire was developed, based on the analysis of the interviews of the first trip, for the survey on the second trip. The result of this survey was also used for cross-examining all the long interviews once the related analyses were completed. Of the 18 questionnaires distributed, 11 were completed and returned. Two more long interviews were also conducted on this trip, with unstructured questionnaires similar to those of the first trip. The long-interview participants were considered 'co-researchers' rather than 'subjects' as a phenomenological study aims to establish a supportive context in which people can build on each other's insights (Seamon, 1979).

A general consensus is that the number of co-researchers vary depending on the nature of the study – between 1 and 10 are recommended (Kaam, 1966; Osborne, 1990; Sousa, 2014). However, it is also imperative to keep in mind that sampling theory is not the basis for selecting the co-researchers since phenomenological research aims not to statistically generalise but to understand human experience (Rosie *et al.*, 2015; Valle & Halling, 1989). All seven co-researchers were selected based on their willingness to participate and that their profiles met the sampling criteria for choosing them (Kaam, 1966; Moser & Korstjens, 2018). The long interviews were audio taped using a portable cassette recorder, and subsequently transcribed. The results of the analyses were cross-examined using the results of the second questionnaire survey, with information such as the tourists' purposes of overseas tours, their expectation of the tour guides, their pre- and post-trip knowledge of the destinations, as well as their expectations of the specific trips to and within China. The third questionnaire was then developed based on the cross-examination results, to collect all their post-tour reflections with regard to China. This instrument was sent to all seven participants a few months after the trips. Six participants completed the questionnaire. The analyses of the long interviews were cross-examined again by the results of the third questionnaire survey. Because the five co-researchers of the first trip did not answer the second questionnaire, they were also asked to answer it along with the third questionnaire. The purpose of the two questionnaires was to ensure, to a certain degree, that their perceptions of the tourism experience were representative.

Initial contacts with the co-researchers began as soon as the tour groups arrived in China. The first round of questionnaires was distributed to all 39 tourists by sliding them underneath door of the hotel room so as not to disturb them after a long-haul international flight. The questionnaire clearly stated the purpose of the research, helping the tourists decide whether to participate as co-researchers. It also helped the researcher to establish rapport, develop participation interest, gain co-researchers' consent for interviews and collect their demographic information. The purpose of developing interest and establishing rapport was to minimise constrained answers. If the participants perceived the researcher to be an expert, they might be reluctant to offer their opinions (Kearns *et al.*, 2017; Riley, 1993). 'Self-exposure procedures' (Bernard, 2006; Lincoln & Guba, 1985) were applied to reduce the power differentials between them. The researcher attempted to share personal background information with the participants in 'idle chatter' (Douglas, 1985; Morishita, 2001; Riley, 1993) to initiate conversations (Figure 1.1). Simple descriptive questions were utilised to help the co-researchers draw on information from their past experiences (e.g. their own life history and previous tourist trips) before talking about other issues. 'Why' questions were generally avoided to reduce protective responses, as they imply a request for a justification of responses and engender defensiveness (Patton, 1990; Pollio *et al.*, 1997).

Data analysis

First, the long interview transcripts were read through to gain a feel for the language and the perspectives of the co-researchers. In this process,

Figure 1.1 Chatting with co-researchers (Source: Author)

their life world was revealed through a comparison of the conversations with their demographic information. Preliminary understandings of their experiences were then written in the margins. The transcripts were then re-read line for line. Significant statements that referred specifically to events, such as 'learning in travel' or 'a cultural shock incident', were selected. Those statements were re-read afterwards with the intention to fragment the raw data into component parts. Each discrete part was eventually categorised using a label to describe each component. For example, the label 'l + t' represented 'learning in travel' and the label 'cse' represented 'a cultural shock experience'. Labels were grouped into sets of relationships, and subsumed under conceptual categories when they assisted in contributing to one. All labels were compared for possible connections and combinations. For example, the labels 'cse' and 'pls' representing 'cultural shock experience' and 'problems of learning situation' were combined and relabelled 'ltp' representing 'learning in travel and its problems'.

The conceptual categories were compared across each individual interview to sort out the core ones that had a pervasive presence in the data (Strauss, 1987; Strauss & Corbin, 1990). Those core categories were again compared across all the long interviews to identify common themes in the data (Colaizzi, 1978; Rosie *et al.*, 2015). For example, themes such as 'travel leads to personal growth and development', 'interaction and processing of experience' and 'establishing a bond with the "other"' were identified because they occurred regularly across all the interviews. In order to increase the research rigor, those common themes were cross-examined using the results of the second and third questionnaire surveys. This exercise not only led to some deduction and combination of existing themes, but it also helped to produce some new themes. The common themes were analysed after cross-examination and placed into natural groups to integrate the long interviews into a thorough description of the individual tourism experience.

Member check

Scholars argue that it should be the co-researchers who determine whether the analysis is an accurate reflection of their conversations (Lincoln & Guba, 1985). Accordingly, participants should perform a 'member check' (Birt *et al.*, 2016; Colaizzi, 1978) to examine the correctness of the description of their experiences. For this specific research, however, only five of the seven transcript analyses underwent a 'member check' because the other two co-researchers stated clearly after the interviews that they would rather read the final research report if possible. The 'member check' was done through either email or regular post mail communications. If it was performed through the postal service, self-addressed and postage paid envelopes were provided to ensure that the final 're-interview' files were returned.

Research Findings

The research revealed that the co-researchers had certain similar experiences despite some differences. Table 1.1 shows the similarities and differences by suggesting the frequency of an experience with corresponding numbers. For example, '7' indicates an experience for all the participants and '4' indicates an experience for four of them. The data were cross-examined using the results of the second and third surveys, thereby producing a summary of the shared experiences (Table 1.2). However, it is worth noting that the long interviews and the questionnaires did not limit the investigation of the co-researchers' experiences in China. The ultimate purpose of this research was not to generalise but to understand by examining specific tourists' views of their past, present and future tourism experiences. In this way, phenomenological description respects understanding not only their experiences on the two trips, but also their reflections/expectations of past/future trips.

It is critical to note that for two reasons the researcher might not be able to conveniently conceal the personal behind the professional in compiling this description. First, the description was based on data collected from the long interviews, each of which had been a dialogical process structured by both the interviewee and the researcher. Second, the interview questions were unstructured and open-ended. To describe the conversations would require a linguistic transformation of the conversations into more precise descriptive terms. This would include shifting

Table 1.1 Differences and similarities in co-researchers' experiences

Tourism Experience	Co-1	Co-2	Co-3	Co-4	Co-5	Co-6	Co-7
Global importance and personal need	7	7	7	7	7	7	7
Mind broadening through tourism	7	7	7	7	7	7	7
Understanding the difference	7	7	7	7	7	7	7
Need for mutual understanding	7	7	7	7	7	7	7
Toward a deep sense of morality	7	7	7	7	7	7	7
Establishing purpose of travel	7	7	7	7	7	7	7
Seeing the self through the "other"	6	6	6	6	6		6
Seeing things beyond the surface	6	6		6	6	6	6
Being selective in taking packages	5	5	5		5		5
Getting out of the tourist trap	5	5		5	5	5	
Education for an individual	4			4	4		4
Satisfaction enhanced by learning		4	4	4	4		
Positive effect on personal career				4	4	4	4
Learning in travel and its problems		3	3	3			
Frustrating and stimulating learning	3	3	3				
Desire to learn more with satisfaction	3	3	3				
Enjoying the thrilling moments	3	3					3
Positive experience enhances satisfaction	3	3			3		
Negative experience affects future travels		2	2				
Mixed feeling toward change			2	2			
Instant gratification in travel					1		
Seeing and reading history						1	
Understanding the present in history							1
Entertaining and purposeless learning			1				
Frustrating and valuable culture shock				1			
Immediate contact with the reality				1			
Chance of testing information				1			
Field of applying knowledge					1		
Limitation of formal learning		2				2	
Tourism's dependence on education					1		
Egocentric motivation of travel				1			
Using professional knowledge in travel						1	
Scratching the surface to learn				1			
Knowledge is the power				1			
Seeking authentic experiences						1	
A guide is the ticket to a good trip			1				
Using the tourism infrastructure						1	
Difference of a tourist and a traveler						1	
Being critical of negative things				1			
Meeting strangers and making friends							1
Facing disappointment							1

from the subjects' original language given in the raw data to descriptions in the words of the investigator (Kaam, 1966; Welch & Piekkari, 2006). 'Linguistic transformation is carried out by means of the ordinary human capacity to understand the meaning of statements' (Polkinghorne, 1989: 52). Therefore, the inter-subjectivity of the long interviews and their verbal descriptions would require both the readers and the investigator to be aware that the latter's gender, age, race/ethnicity and biography

Table 1.2 Tourism experience for transforming a social being*

The Significance of Tourism Experience
Travel Leads to Personal Growth and Development
Global importance and personal need
Mind broadening through travel
Education for an individual
Satisfaction enhanced by learning
Positive effect on personal career
The Transformation of a Social Being
Interaction and Processing of Experience
Establishing purpose of travel
Seeing things beyond the surface
Being selective in taking packages
Getting out of the tourist trap
Establishing a Bond with the "Other"
Understanding the difference
Need for mutual understanding
Toward a deep sense of morality
Seeing the self through the "other"

*Co-researchers' shared experiences.

could position the researcher in the description. All of these might inhibit or enable insights throughout this research. Table 1.2 indicates the co-researchers' shared experiences. It is hoped that an understanding of their shared experiences will provide insights for both tourism research and business practices.

Meaning of tourism experience

The co-researchers were motivated to travel in the hope of having a good time. They wanted to see more of the world, learn about other cultures and experience places they had read about in books. Driven by such motivations, they found themselves, consciously or subconsciously, engaged in learning situations during their travels. Although the learning might not be pleasant, as it was experiential and occurred incidentally, they did feel afterwards that experiential learning of their travels opened their eyes to a broader picture of the world. A tourism experience is thus significant for an individual, particularly in an era when cultures and economies are increasingly integrating into a globalised system.

The co-researchers agreed that the fundamental appeal of the tourism experience is the possibility of accessing distant places, thereby learning in person basic knowledge about the nature of human life on a global scale. Co-researcher 2 pointed out:

> Travel is about learning the very basics of life. We are all humans, we share common concerns, and we have the same basic needs. We need to know others' needs and wants to gain insights for our own life. (Author's direct interview quotation)

Her view, affirmed by Co-researcher 5, is that 'travel is not only about the interaction of peoples, but also the integration of world nations', which is 'the practical weapon to stop wars' in the opinion of Co-researcher 4. The global importance of cultures makes the tourism experience significant for personal growth and development, as explained by Co-researcher 6 when explaining his three reasons for joining the tour of China:

> My first reason is intellectual, because it's hard to think of yourself as an educated person if you know nothing of the Chinese thinking. The second is political. I have been critical of Western capitalism most of my adult life. This is not to say that I think any of the existing forms of socialism is perfect, but I've found great Western inspiration for challenging injustice by looking to the experiments in socialism. The third is economic. The changes from planned economy to market economy, and the move of many Chinese people from China to Canada, are economic processes. I think it's hard to understand the economy, certainly of Vancouver, or Toronto, without understanding how Chinese people do business. [Author's direct interview quotation]

It seems that those participants were attempting to gain a sense of harmony with a life that is increasingly integrated into the global system. They chose travel as a means to achieve this harmony because, as pointed out by Co-researcher 3, it is 'a great education for an individual'. This opinion is also reflected in the view of Ryan (1997) that leisure tourism possesses the potential for cathartic experiences because it is an entertainment business involved with learning. The experiences may not have life-changing effects, but do nevertheless have certain significance for tourists' personal growth and development.

Transformation of a social being

The following statement on the social transformations of tourism may help further understand the co-researchers' views on the significance of tourism experiences:

> Closely examined, the social transformations of travel strongly suggest that social being in general derives from nothing more than the mutual

> identifications, categorisations, and recognitions in which people normally engage. (Leed, 1991: 263–264)

This quote suggests that these transformations are closely related to the origin of identity, the ways in which an individual's self is defined. Namely, without 'other' there would be no 'self' and identity is established with a 'mirror'. This 'mirror' frequently appeared in the co-researchers' experiences whereas their identities were reflected in the recognition and observation between 'self' and 'other', as well as in their interacting and processing of experiences before/after a specific tourist trip.

All the co-researchers attempted to interact and process their satisfaction with past experiences, which worked as a 'mirror' to reflect the 'yes' or 'no' for their future experiences, with 'yes' and 'no' representing past experiences to be repeated or avoided in their future travels. As experienced world travellers, they had already sorted out the 'yes' and 'no' from past experiences, evident in that they all wanted to avoid the 'inauthentic' experience manipulated by the tourism business – the so-called 'tourist trap'. However, they understood that in the age of mass tourism it was unrealistic to completely avoid 'the tourist trap', having to use the tourism infrastructures for individual needs, as Co-researcher 1 explained:

> On a package tour, the first thing I get upon arrival will be a hotel room. It's like a home. Then I can explore gradually, further and further, because I can continually come back to this home base... the more time you have to spend finding things out, you waste your time. If someone else helps you, you have the information and you can do it. (Author's direct interview quotation)

If avoiding 'the tourist trap' to 'explore further and further' into a destination is part of a satisfactory tourism experience, the deeper knowledge obtained about the 'other' may be a means of changing selves. This self-changing effect is reflected in their frequently rediscovering 'self', while seeing the 'other' during their travels.

Establishing a bond with 'other'

Co-researcher 1 had to re-adjust her view of the 'other' which had been constructed by her home media after she realised that the portrayal of the 'other' had caused her to misjudge the local people. Finding selfhood in the recognition of others was 'a saving grace' for Co-researcher 5. He believed in travel's contribution to making successful tourists into actors, writers and poets. Co-researcher 4 felt that he should not take his wealthier life for granted after seeing the poverty in developing countries. By seeing the life of the 'other', Co-researcher 2 felt that she could draw insights for her own life. Co-researcher 3 often compared what he had seen during his travel with familiar things at home. Co-researcher 1 believed that one could pick up pieces of knowledge on a trip, and put them into one's knowledge box to make oneself a better person.

The above performative manifestations show the significance of understanding the differences of the 'other' for the co-researchers' tourism experiences. Co-researcher 6, who was a teacher and researcher of adult learning, believed that this understanding would contribute to world peace and harmony:

> Whatever problems tourism has, if we look at the informal travel experiential learning, we find it contributes to life-long learning that contributes to global peace and harmony. If I'm open to learning, your difference becomes a resource. Otherwise, your difference may be a threat to me. (Author's direct interview quotation)

Conclusion

The findings of this research provide an understanding of the co-researchers' shared experiences. Namely, travel is joined with other dimensions of living – the socioeconomic, interpersonal and spiritual worlds that, in sum, comprise an individual's geographical consciousness. Understanding this relationship between the tourism experience and geographical consciousness points to wider patterns of meaning with regard to tourists' behavioural and experiential relationships with places, particularly patterns of how the 'body subject' relates to the 'lived space'. According to Seamon (1979), underlying people's everyday movements is an intentional bodily force manifested automatically but sensitively, such as feet carefully walking to a destination, or an arm reaching for an object. Seamon (2013) terms this as intentional bodily force or body subject. Because of this, an individual knows at any particular moment how to relate a normal daily experience to familiar objects, places and environments. Thus, it is necessary to conceive of a lived space: a space organised in terms of a corporeal scheme and constituted by means of bodily movements such as actions in response to specific situations (Ashton, 2014; Zaner, 1971). For example, a co-researcher couple wanted some 'independence' on their 'dependent' package tour of China (see Figure 1.2):

> We would like to have some afternoons or evenings off, say, walking from our hotel to just anywhere, and find somewhere. We would like to go to local bakeries and look in, wandering what this is and say, 'let's buy it and find out what the taste is'. That to us, is travel. (Author's direct interview quotation)

However, they might not have attempted to be 'independent' until they had developed a sense of the lived space of China. Without such

Figure 1.2 A 'cultural tourism' activity of a co-researcher couple (Source: Author)

sense, they would not have the intentional bodily force – the body subject that would guide them, automatically but sensitively, to explore China independently. As one of the couple conveyed previously, the benefit of an organised tour was having help from someone else. Otherwise, one would spend more time finding things out, and wasting one's time. Therefore, to study a tourist's experience of a destination, it is essential to recognise that this is grounded first of all in the body of the tourist, in her/his geographical consciousness. This recognition has significant implications for tourism research.

Theoretical implication

Recognising the role of geographical consciousness in the tourism experience may help understand tourists' 'freedom of choice' and 'cognitive bias' during travel. Tourism is, after all, a social, cultural and economic phenomenon, as well as a personal experience (Amoiradis *et al.*, 2021; Li & Hinch, 1997; Squire, 1994a). The general trend of cultural change in a given society affects tourist understanding of the nature of tourism. However, the experience of tourism requires, first of all, a physical body. While a society creates a context for a tourist to experience, the diversity of views held within it attributes an importance to the tour:

> Social morals are individually interpreted, so the experience of being a tourist is one that engages all the senses... To understand the tourist experience requires not only a consideration of perceived needs and

> actualities of the beach and the great outdoors, but also the noise and dins of the disco and the sweat of the massage parlour. (Ryan, 1997: 25)

The points that Ryan makes in the above quotation are that the experience may be represented by diverse behaviours, because social contexts are pluralistic in nature, offering tourists freedom of choice – many opportunities for expressing different behaviours.

For example, Co-researchers 1 and 6 would like to have more 'independent' activities on a 'dependent' package tour of China, such as going on their own to the local bakeries, shops and restaurants. On the other hand, elderly people, such as Co-researchers 4 and 7, would visit local galleries, teahouses, open markets and schools with the tour group at a leisurely pace. Both may occur, albeit at different times, and there seems little in common between the two types of behaviour. One contains the potential for aggression and the other is a re-enactment of a more genteel past. However, both sets of behaviour arise from a need for socialisation with the locals to have an 'authentic' experience. Evidently, tourists' needs are few, while their expressions of those needs are diverse (Ryan, 1997). It is the freedom of choice that confuses, and leads to a cognitive bias excluding other strata of the tourism experience that can also be 'authentic'. The neglect of freedom of choice and cognitive bias is probably an important reason for some misleading notions about authenticity.

Therefore, an unresolved issue exists in tourism research – the nature of tourism authenticity (Selwyn, 1996). On the one hand, MacCannell (1973) asserts that tourists' search for a sense of authenticity to reclaim that which has been lost by an isolating and fracturing postmodern life. On the other hand, Cohen (1989) develops this sense in the claim that, because 'educated drifters' share senses of alienation, they are drawn to the authentic. Both senses, however, represent a similar notion that authenticity is the luxury of the 'elite' tourists' constant feeling of alienation from 'self' and dream about the traditional, primitive and exotic lifestyles of the 'other'. Even if the traditional no longer exists, changing with the penetration of global tourism, there still needs to be 'staged authenticity' that provides tourists with 'pseudo-events' (MacCannell, 1973) representing the exotic 'other'. The 'other' still needs to present a self-image of being 'primitive' to the explorer type of tourist (Cohen, 1989). Considering the cognitive factors that cause freedom of choice and cognitive bias, it is questionable if tourism authenticity could also be non-traditional, 'civilised' and constantly changing.

Freedom arises out of limitation (Seamon, 1979, 2013). A person may want to get out of 'the tourist trap' in order to have the 'authentic experience' they desire. However, they may also have to depend on the tourism economy to achieve those wants:

> Without the tourist infrastructure, I wouldn't be able to find my way. It's the language barrier, culture complexity, and short time. But I need some space away from the group. By the end of the day, I would walk out of the hotel to see the city by myself. When I started to walk back yesterday, I could see a policeman coming home taking off his belt. I could see a soldier coming home taking off his hat. I could see a shop assistant coming home putting a jacket over her uniform dress. I felt that I was where the local people lived, which was nice. (Author's direct interview quotation)

The above quote is a co-researcher's description of one of his daily tourist activities on the 'dependent' package tour of China. The body subject (his intentional bodily force for exploring the destination independently) and the cognition (the realisation of potential difficulties in using the force) coexist in this experience. He wanted to be independent, yet was not capable of coping with the 'language barrier, culture complexity, and short time' on his own. In the end, he had to revert to the regular route – joining a package tour and making the tourism infrastructure for his individual use.

The above example also suggests that cognition or cognitive factors in the experience should be balanced with the general routine operations of the tourism industry, when a tourist tries to achieve travel satisfaction. Cognition in general terms is a true state of understanding a phenomenon (Johnston *et al.*, 1988). In a more specific term, it is an innovative force by which one can do other than what one has done in the past. A tourist can imagine their life being different, and then work to make that image a reality (Seamon, 1979, 2013). However, thinking about doing differently does not necessarily lead to different actions. Cognition may encourage a tourist to imagine an 'authentic' experience, while the same cognitive factor may lead the tourist to understand that, in reality, their imagined 'authentic experience' is hard to achieve. Thus, a tourist's cognition has to be balanced with the regular operations of the tourism industry in order to achieve satisfactory experiences. A means of weighing the routine operations of the industry against one's cognitive factors would also help to gain a better balance between the tourist's dream of authenticity – the exotic, remote and primitive destination – and the rights of the destination to develop and change with increasing global tourism. Regarding this balance, one co-researcher shared some insightful opinions:

> Otherwise, we take the traditional China. We put a plastic bubble over it, preserving it perfectly for tourists to look at, but it basically is not the real China now… We should be interested in something unique, but for it to be truly unique, it is not a product or a souvenir. It is an authentic human experience. For it to be genuine, real, authentic human experience, it will change.

Practical insights

This research demonstrates that both the tourism experience and tourist cognition – geographical consciousness – are joined. Cognition is significant not only for achieving a satisfactory travel experience, but also for establishing a bond between 'self' (tourist) and 'other' (host), thereby contributing to personal growth and development. It is essential, therefore, that the tourism experience is understood not merely as a commercial activity. Rather, it is an ideological framing of history, nature and tradition (MacCannell, 1992), constituting experiential learning that takes place in the process of geographical consciousness. Understanding this has the following practical implications for the tourism industry.

First, geographical consciousness manifests in tourism experiences as the spatial and temporal bond of people–place between tourists and destination. It is represented by the fact that the routine operations of the industry, the social reality of the destination and the cognition of tourists, all influence the tourism experience. In this vein, an authentic tourism experience does not necessarily mean the search for an exotic, remote and primitive 'Shangri-La', but perhaps a desire for experiencing the vibrant life of a destination, including its development and change. This should be appreciated by the tourism industry in resolving the dilemma between tradition and the modernisation of tourism destinations.

Second, this research shows that the tourism experience represents a discourse of learning, that knowledge can be obtained through understanding and collecting differences about 'other' people and places. Those differences are novel and strange and are experienced by tourists in their quest for ways to satisfy a wide range of personal needs, from pleasure to the search for meaning (Cohen, 1979). The tourism industry should appreciate that the travel business contains a cultural capital that can be collected, and that it possesses a potential to educate tourists while entertaining them. As tourists have to rely on the tourism industry to collect the cultural capital, the industry can encourage or even manipulate the process, through its representatives, such as travel agents and tour guides, to enhance tourists' accessibility to the capital. Tourism practitioners thus need to situate tourism in other realms, i.e. adult education, recommending more alternatives such as educational tours, for the benefit of both the tourists and the travel business.

Third, from the co-researchers' shared experiences, it is evident that the educational dimension of tourism may lead them, at different points, to move along a continuum toward an awareness of a personal bond with a destination. That is, learning through a tourism experience begins with the anticipation and planning of a specific trip. Upon arrival at the destination, the tourist starts their on-site experiences (e.g. sightseeing, dining, souvenirs shopping). When the tourist returns home after the specific package tour, they recollect the impressions of the experience. This process is influenced, and may be manipulated, by the travel industry, because it is this entity that facilitates the experience. In the meantime, tourists' judgements of the valuable aspects of the experiences are filtered through a process of recollection and evaluation. Those valuable aspects are parts of the past experiences that may be repeated in future travels. The whole process of the tourism experience – anticipating and planning, travelling to the destination, the on-site experience, travelling back, and recollecting and evaluating, may move in two directions. One may be positive and the other negative towards tourism development, depending on the performance of the travel industry that facilitates the experience. If the industry tries merely to attract a large market for economic benefits, neglecting such a potential environment associated with tourism expansion, may lead this process in a negative direction. However, if the industry is truly concerned about the overall well-being of its business, and it attempts to engender that concern in tourists through appropriate planning and management, this may lead the process in a positive direction. In either direction, tourists' judgements, based on their recollection and evaluation of past experiences, would provide another starting point for a new tourism experience.

The above illustration of the tourism experience continuum also indicates that a potential personal bond with the destination may help tourism business sectors to understand, holistically, the spectrum of the experience in contemporary society whose conditions are leading people away from traditional values to new or alternative ones, particularly to a changed postmodern condition of knowledge (Kalinowski & Weiler, 1992; Lyotard, 1984). An important characteristic of the postmodern condition is valuing knowledge in terms of its 'performativity', which emphasises the optimisation of an efficient performance by using that knowledge (Chalise, 2019; Harvey, 1989, 1996). This valuing of it and of its performativity suggest that there should be a co-implication of contemporary discourses of individualistic learner-centeredness and trends towards the marketisation of learning opportunities (Falk *et al.*, 2012; Usher *et al.*, 1997). One would thus argue, from a sociocultural and psychological point of view, that learning is a characteristic of all people involved in the tourism industry, including tourists, operational personnel and people at destinations. The industry needs to have a holistic view of the tourism industry when evaluating, estimating and determining tourist wants and needs in order to develop new programmes. By doing so, the industry may develop appropriate products both for tourists' expectations of satisfactory experiences and for those, such as travel agents and tour guides, who facilitate those experiences.

2 Orientalism Revisited: Ethnic Tourism of China versus Canada

Introduction

This chapter presents comparative case analyses of ethnic tourism between China and Canada. Through a critical review of Edward Said's (1978) *Orientalism*, attention is directed towards the development processes, social/cultural issues and tourism development strategies of two specific cases. In doing so, the author argues that Said reconfigures the historical construction of European and Euro-American discourses about Near-East civilisations, cultures and peoples. However, his focal point of view is grounded in the examination of American, English and French colonialism and literature of the cultures of the Near East and Palestine, with scant attention given to China and East Asia. Influenced by Said's *Orientalism*, traditional ethnic tourism tends to create a stage that displays some exotic and quaint cultures and customs of 'others' for the tourist gaze. Consequently, existing ethnic tourism may have overlooked the delicate balance between the economic and social-cultural interests of the ethnic groups involved in the process of tourism development. Considering the different social, cultural and economic backgrounds, this specific research sought to uncover the extent to which those interests would manifest differently between China and Canada, among the ethnic groups involved in the tourism industry.

Decades ago, scholars predicted a promising future for this trend in world tourism development – among the millions of tourists, increasing numbers would seek cultural experiences of other ethnic groups (Picard & Wood, 1997). Such a phenomenon has substantial implications for the development, management and marketing of tourism destinations. Tourism marketing has thus become more culturally oriented, and the chosen attractions for many tourism destinations tend to be ethnic cultural heritage resources. Nonetheless, developing ethnic tourism is not as simple as it might first appear. Through the years, many scholars have raised a series of questions for the tourism industry regarding ethnic tourism. What features should be selected? What is attractive to a chosen market? And just as important, what would be objectionable and to whom?

(Boniface & Fowler, 1993; Chistyakova, 2020; Hitchcock, 1999; Li, 2011; Zhuang *et al.*, 2017).

Ethnic tourism traditionally caters to tourists' desires to learn about, witness and experience faraway cultures of the so-called 'Orient' or 'other'. The fact that tourists place a high value on opportunities to experience 'other' cultures is demonstrated by their willingness to allocate substantial resources in the form of money, time and energy. While ethnic destinations strive to turn those expenditures into local revenue and income, some critics warn that tourism may actually destroy the ethnic hosts' cultures (Turner & Ash, 1975). This criticism, however, contrasts with the argument that tourists' satisfactory experience could be matched by tangible rewards for the ethnic hosts, which are primarily of an economic nature (Mill & Morrison, 1992). The reality is that a variety of social/cultural, economic and environmental impacts accrue to the ethnic tourism destination, some of which are positive and some are negative. Ethnic tourism destinations must grapple with the difficulty of fostering the positive impacts while discouraging the negative impacts, in order to run a sustainable tourism business. The sensitive nature of ethnic cultural resources thus represents a challenge to the development of sustainable ethnic tourism.

Literature Review

The connotation of so-called faraway cultures in association with traditional ethnic tourism is very much a manifestation of Edward Said's configuration of the 'Orient'. Accordingly, this chapter begins with a critical review of the 'Orientalism' theory, utilising it as a lens to theorise the key concepts of ethnic tourism and sustainable development, which underpin the very research that formulates the author's discussion of the phenomena in question.

'Orient', ethnic tourism and sustainable development

'Orient' is a European invention constructed to represent the 'other'. According to Said (1978), an orientalist is a practitioner who makes systematic studies of or teaches about the Orient, and orientalism is an ideology based on an ontological and epistemological distinction between the East and the West. Scholars, especially those in the disciplines of humanities and social sciences, have accepted this distinction as the standard for their literary creations and social and political descriptions. Orientalism can be discussed and analysed as an instrumental system for treating the Orient in different ways (Shabanirad & Marandi, 2015) – 'making statements about it, authorising views of it, describing it, teaching it and above all, ruling over it' (Said, 1978: 43).

Said maintains that the Western discourse of Orientalism enables the enormous systematic discipline by which European culture is able to manage and even produce the Orient politically, sociologically, militarily, ideologically, scientifically and imaginatively. Literary and historical texts are constructed within the political contexts of colonialism, creating a discourse about the East as the cultural 'other'. In line with Foucault's (1980, 1990) thought, Said's 'Orientalism' is not only an academic subject that teaches knowledge about the Orient, but also a style of thought that draws a clear distinction and an opposing but no less enduring Western view. Accordingly, the Orient is regarded as a social construct, just like Foucault's idea of the human construct. However, no Western discourse, literary or political, about a nation, culture and people can be the passive within the context of imperialism and colonisation because, to a certain extent, all participants maintain a cultural sovereignty (Zhang, 2019).

The history of Orientalism has both an internal consistency and a highly articulated set of relationships to the dominant culture surrounding it. Literary and historical texts are constructed within these political contexts, creating a discourse about the East as the cultural 'other' to the West. As a result, Oriental history for Western philosophers of history was useful in portraying a region of great age and what had to be left behind. Orientalism accordingly initiates a new kind of study of colonialism and postcolonialism. Said believes that the representations of the 'Orient' in European literary texts, travelogues and other writings have contributed to the creation of a dichotomy between Europe and its particular 'difference' – a dichotomy central to the creation of Euro-centralism, as well as to the maintenance and extension of European hegemony in other lands. This dichotomy is also crucial to the self-establishment of Europe. If colonised people are irrational, Europeans are rational; if the former are barbaric, sensual and lazy, Europe is civilised, with its sexual appetites under control and its dominant ethic of diligence; if the Orient is static, Europe can be seen as developing and marching ahead. This dialectic between self and other has been greatly influential in subsequent studies of colonial discourse, including tourism research on the relationship between Oriental hosts and Western guests (Ooi, 2005; Shim *et al.*, 2015; Smith, 1989).

The tourism literature often makes reference to the 'quaint' or 'exotic' customs of 'other' peoples, as manifested in so-called ethnic tourism. Usually, ethnic tourism involves some form of direct experience with non-Western or indigenous cultures and environments. According to Figueroa-Domecq *et al.* (2020), Moscardo and Pearce (1999) and Nielsen and Wilson (2012), the human element is important in ethnic tourism, whether it be visiting ethnic homes and villages; observing and participating in local ethnic customs, ceremonies, rituals, dances and other traditional activities; or studying or purchasing indigenous artefacts, arts and crafts. Scholars have made a distinction between ethnic tourism and general cultural tourism. While cultural tourism involves exposure to a culture in an indirect way, ethnic tourism involves first-hand experience with the practices of another culture to provide tourists with a more 'intimate' and 'authentic' experience (Chistyakova, 2020; Greenwood, 1982; Klieger, 1990; Wood, 1984). Other scholars believe that consumers of ethnic tourism exhibit a common desire for authenticity, immersion in the cultural and/or physical environment and the pursuit of environmental and experiential quality in an attempt to experience novelty and uniqueness as part of the travel experience (Fraser, 2020; Weiler & Hall, 1992). In general, the practice of ethnic tourism aims to offer a transient opportunity for tourists to capture some of the 'intimacy' and 'authenticity', and the novelty of experiencing different 'other' cultures.

Notwithstanding its cultural context, ethnic tourism is an economic phenomenon. It exists due to the forces of demand and supply. More specifically, ethnic tourism exists due to the scarcity and uneven distribution of ethnic cultural resources (Figure 2.1). The market is made up of people who seek a unique ethnic cultural experience and who demand certain resources that they feel will facilitate their experience. However, like most other types of economic activity, further processing of basic resources is required. While the raw cultural resources in a destination may provide the context and environment for tourism, further refinement is necessary to produce a tangible product appropriate for tourist consumption. The tourism literature commonly refers to this process as the commodification of culture (Graburn, 1989; Hitchcock, 1999). However, concerns have been raised regarding the negative connotations attached to the process, based on arguments that commodification affects the authenticity of a culture. Nonetheless, it can be countered that commodification may represent a mechanism to protect cultural resources. Destinations that selectively transform cultural resources into tangible

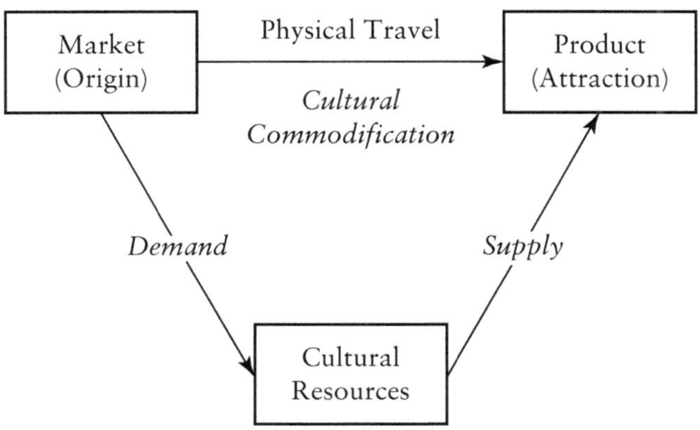

Figure 2.1 Economic perspective of ethnic tourism (Source: Author)

products not only facilitate the exchange of this cultural experience for financial return, but also have the potential to create a situation in which the destinations can promote sustainable development through careful management of resources (Craik, 1991). While the market's original demand is focused on the resource, it also tends to consume the resulting processed product. The physical travel of tourists occurs between the location of the market (origin) and the location of the product (destination). To the extent that the production and distribution of the product is controlled by the destination (temporal and spatial), the ethnic hosts have the opportunity to protect their cultural resources.

The above illustration suggests that the commodification of culture in the context of ethnic tourism is not necessarily a destructive process. Nonetheless, the exploitation of ethnic tourism is not a panacea for development. The negative impacts of ethnic tourism can be significant, causing local inflation, the denigration of sacred sites and environmental pollution (Ishii, 2012; Mathieson & Wall, 1982; Ryan, 1991; Swain, 1989). While it is convenient to categorise those impacts into specific realms (Figure 2.2), it should be appreciated that those realms are interconnected and that the tourism system as a whole is not closed to outside influences. The manifestation of tourism impacts on a destination is therefore complex. Development strategies represent an attempt to ensure a net positive balance by maximising the benefits and minimising the costs. This, however, is not an easy task since decisions must be made and strategies developed in a dynamic environment involving many interrelated variables. Therefore, ethnic tourism is a controversial phenomenon, raising fundamental questions with regard to sustainable development in the process of ethnic culture commodification. This may create some especially controversial issues if the Western concept of ethnic tourism were applied to the Chinese situation of ethnic minority cultural tourism. The connotation of faraway cultures of ethnic tourism would merely manifest the configuration of the 'Orient', if it does not direct attention to China and East Asia by relying largely on Said's

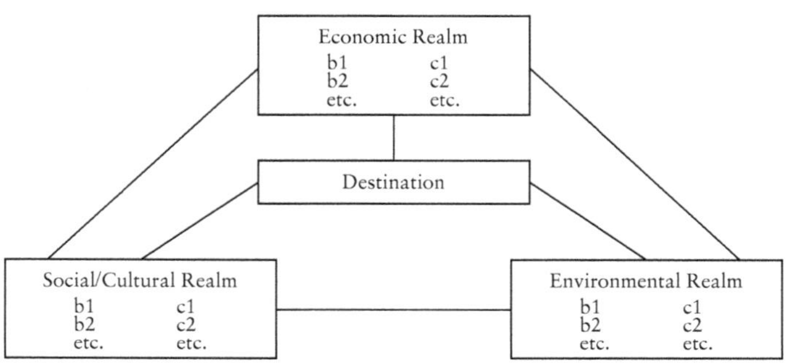

Figure 2.2 Manifestation of ethnic tourism impacts (Source: Author)

perspective of 'Orientalism' that grounds the focal point of view on the examination of Euro-American colonialism and the literature of the cultures of the Near East and Palestine. In so doing, the sustainability of the associated ethnic tourism would be out of the question.

According to the World Commission on Environment and Development (WCED, 1987), sustainable development aims to meet the needs of the present generation without compromising the ability of future generations to meet their own needs. Gunn (2002) believes that sustainable development principles are applicable to tourism for two reasons. First, the basic premise of sustainable development is not only to make growth possible but also that the growth is positive – to provide net social and economic benefits. Second, sustainable development aims to generate a stable and undisturbed resource base. In other words, sustainability is based on the balance between stability and change. Such a balance is possible to achieve in ethnic tourism, considering the characteristics of the tourism system wherein benefits and costs coexist (Figure 2.2). Ideally, tourism practices should provide tourists with a rich, accurate and entertaining understanding of cultural resources (Gunn, 2002). For instance, one of the reasons why an ethnic tourism site is established is to relieve the tourist pressure on the real environment. The key to success is that the facility within the site should evoke a fascinating and memorable tourist experience, while preventing degradation of the cultural resources of the 'real environment'.

In sum, it is at a tourism site, where the tourism transaction occurs, that the specific issue of balancing resource protection with development transform from policy to action. Sustainability is, however, not inherent to ethnic tourism. 'Tourism is dynamic and not a static phenomenon' (Butler, 1993: 29). Similar to other industries, tourism is an agent of development and change. It is consumptive like other industries, and the level of consumption is determined by how tourism is developed (Woodley, 1993). At certain levels of consumption and with careful planning, tourism may be able to operate in a sustainable fashion. Controlling the level and style of development over the long term presents substantial challenges, since potential impacts must be addressed throughout the planning and management process. Developers need not consider this action a constraint but rather a necessity for doing business. What they do need to ponder is how to maximise the social/cultural and economic benefits while minimising the associated costs.

Research Methods

Swain's (1989) conceptual model of indigenous tourism development was utilised to develop a theoretical framework for this research. The model defines the units of analysis as the nation-state, the tourism industry and the ethnic group associated with ethic tourism development. The

Table 2.1 Conceptual model of ethnic tourism development

ASPECTS OF ANALYSIS	*Sphere of influences*		
	NATION-STATE	***TOURISM***	***ETHNIC GROUP***
Development process	Ethnic autonomy	Tourism marketing	Social/cultural response
Paradoxical issues	Nation-state regulation vs ethnic control	Culture freezing vs evolution	Cultural pluralism vs integration
Strategies for development	As above	As above	As above

Adapted from Swain (1989).

model was modified to appropriately fit the analyses of ethnic tourism in the selected case studies (Table 2.1). It illustrates the interrelationship of the primary ethnic groups in tourism development with the characteristics of the development processes, paradoxical issues encountered and strategies adopted for sustainable development.

Characteristics of development process

One critical characteristic of ethnic tourism development is the actual control an ethnic minority group can exert over the process. If the group has legally recognised power in determining local use of the national infrastructure (e.g. education, communication, transportation and health systems) and the exploitation of natural resources, it is likely to play a strong role in its own tourism development. The cooperation of local, national and international concerns is a central issue in this kind of tourism development. Political autonomy plays a key role in this development process, which benefits ethnic minority groups. The key questions are: (1) Is sufficient ethnic control exercised? (2) How should ethnic cultures be marketed? (3) What social/cultural responses are manifested toward tourism? (4) What are the prospects for future development?

Paradoxical issues

Paradoxes of ethnic tourism occur due to the inherent contradictions between conservation and change in the process of development. This paradox is cross-cut by the fact that viable cultures are not static, but evolve over time. A restructuring of political relationships occurs, which places the nation-state in the position of promoter and regulator of cultural forms, while validating ethnic minority people's awareness and legal rights. Tourists often expect the locals to be quaintly non-modern or 'frozen' in time, while the locals cannot avoid social/cultural change under the influence of tourism development. The standardisation of an ethnic culture for tourist consumption encourages ethnic groups themselves to package ethnic goods and lifestyles. As a result, tourism activity accelerates the socioeconomic change of ethnic communities, affecting tourism authenticity. Cultural pluralism is an important asset of ethnic tourism, yet political and economic infrastructures tend to integrate ethnic minority people into the majority society, thereby encouraging the assimilation of individuals as well.

Strategies for development

Within an ethnic group, strong geopolitical organisation and economic incentives to stay 'quaintly' authentic to attract tourists might offer a resolution to those paradoxical issues. The articulation of the nation-state economy, the expansion of the tourism industry and local involvement in the development process will shape the outcomes of ethnic tourism. Therefore, culturally and economically sustainable tourism development must adhere to the local society and make cultural sense. Ethnic people themselves must own the process of local tourism development, while the nation-state may benefit both social/culturally and economically from cultural diversity.

Case study selection and investigation techniques

The Yunnan Folk Cultural Village (YFCV) in China was selected as one of the two case sites for analysis. It is located in the west suburb of Kunming, the capital city of Yunnan Province in southwest China. The site covers approximately 230 acres wherein eight ethnic minority attractions were open to the public when the research was operationalised. Seventeen more attractions were planned to be completed by the end of 1996. The YFCV displays selected aspects of traditional and contemporary ethnic minority cultures in China, including architectural style, performing arts, local festivals and traditional music. The other case site is Wanuskewin Heritage Park (WHP) in Canada, which is located on the banks of the South Saskatchewan River in Saskatoon – a major city of Canada's Saskatchewan province. WHP covers 300 acres of land and, in contrast to the YFCV, remains largely in its natural state, featuring a variety of archaeological relics typical of Northern Plains Indians' cultural heritage. Opened in 1992, WHP offers carefully designed walking trails, outdoor displays and performance areas, a full-scale archaeology laboratory and a state-of-the-art visitor centre that includes a gift shop, a restaurant, two small specially designed theatres, an art gallery and a main gallery. Interpretative themes at WHP are based on the period of the Northern Plains Aboriginal Indian culture prior to contact with European-based cultures.

The aforementioned conceptual framework regarding the interrelationship between the primary ethnic groups in tourism development provides the research of these two cases with a workable scope of sampling and data collection. Accordingly, all data collected were utilised to explore the characteristics of the development process, the paradoxical issues encountered and the strategies for development. The primary data were drawn from the opinions of selected representatives from various participant groups associated with both cases. Semi-structured interviews were conducted based on purposive sampling 'to get the most comprehensive understanding' of the phenomena in question (Ames *et al.*, 2019; Henderson, 1991). This sampling technique consists of 'sampling extreme or deviant cases as well as typical cases. Maximum variation was explored... for particular reasons to sample critical cases or politically important or sensitive cases' (Henderson, 1991: 133). Specific individuals for interview were identified, based on the analysis of printed materials and on recommendations of the chief executive officers of the two case sites, to view the critical issues outlined in the conceptual framework (see Figure 2.2).

A total of 15 interviews were conducted, with at least two representatives from each of the three categories: the nation-state level of tourism planning and management, the case sites proper and the featured ethnic groups. Field notes were made during the research trips to record on-site observations, in order to generate factual descriptions of the study subjects and add to the primary interview data. Secondary data were drawn from a comprehensive review of printed documents including facility management plans, marketing strategies, interpretative programme descriptions, tourism promotional materials and other records of a similar nature. These secondary data were used for interpreting the interviews. A typological approach (Henderson, 1991; Stapley *et al.*, 2022) was applied to data analysis to make possible categorical judgements about patterns and themes, evaluate the existing model and generate substantive theory.

The interviewees were anonymised so that respondents' comments remained confidential (Table 2.2). While much of the printed material

Table 2.2 List of interview participants

China	*Ethnic background*		*Canada*
Government official	Minority	Aboriginal	Government official
Government official	Minority	White	Government official
Tourism researcher	Han	White	Site manager
Site manager	Han	Aboriginal	Attraction guide
Tour group manager	Minority	Aboriginal	Educator
Attraction guide	Minority	Aboriginal	Storyteller
Street vendor	Minority	Aboriginal	Elder
Four backpackers	White		

Note: Participants of ethnic background are termed 'minority' in China and 'aboriginal' in Canada was solicited through correspondence with the case sites, data collection trips were made to both the YFCV and WHP to conduct interviews. Methodological concerns related to language and other cultural barriers were partially addressed in the data collection process by the makeup of the research team that included a Canadian professor who had an active ethnic tourism research programme and the principal investigator himself, who is a Chinese national with work experience in the Chinese tourism industry.

Research Findings

The research uncovered the following challenges and achievements manifested in the processes of ethnic tourism development at both the YFCV and WHP.

Unresolved challenges

The lack of ultimate ethnic control presented the first and perhaps the most common challenge to both cases, despite the different causes. This challenge was manifested in the reality that ethnic involvement in the development processes was merely culture interpretation and operating the tourism programme, with a very limited ethnic role in the planning and management aspects. The second major challenge was that the selected culture programmes might contribute to the creation of incomplete ethnic images. For instance, the YFCV focused its programmes on contemporary ethnic life while the emphasis of the WHP's indigenous cultural presentation was on the historical perspective, which could affect the tourists' gaze on the ethnic hosts, and the self-understanding of the hosts about their own cultures and community identifications. Sometimes, selectively and positively presented programmes for tourism purposes may help certain sacred aspects of an ethnic culture avoid erosion and denigration in the face of mass tourists' intrusion. However, selection is a sensitive process which, if handled inappropriately, may not gain the approval of the affected ethnic communities, according to a Canadian aboriginal educator:

> My culture is my life. They take my culture from hundreds of years ago and put it on display. They don't put anything there about contemporary... no continuation of my culture... We are not presented like other Canadians... We are presented just like the Hollywood stereotype... They continue to abuse us. (Author's direct interview quotation)

In the YFCV, it appeared that the attitudes of the Chinese ethnic minority people were more moderate towards the selected ethnic cultural programmes being presented to tourists. Nonetheless, disagreements could still be identified. For example, an ethnic minority vendor disclosed that her community elders were very unlikely to support the young people working in the tourism sector:

> The old people won't let us go and work inside [the Village] ... They think all these pubs, KTV clubs are not good for us young girls... (Author's direct interview quotation)

In terms of the major issues that emerged, the case studies revealed that both sites had faced similar dilemmas inherent in the process of ethnic tourism development. In spite of the fact that great physical and cultural distances existed between the two cases, they both had to deal with these paradoxical issues, including nation-state regulation versus ethnic control; the tendency to freeze ethnic culture versus cultural evolution; and cultural pluralism versus ethnic integration. Developing appropriate strategies to cope with those issues would, apparently, add more challenges to both the YFCV and WHP in their pursuit of sustainable development.

Explicit and implicit strategies for development were adopted in both cases. However, those strategies did not fully address the inherent issues of ethnic tourism development. Both recognised the importance of building a strong and diversified revenue base in order to finance their respective development and operation, but neither seemed to have taken concrete action to build up an alternative financial base that would provide appropriate ethnic control of the development processes. Appropriate ethnic control is considered crucial to development. Otherwise, economic benefits could be inequitably distributed among industry entrepreneurs and ethnic minority communities. The unequal distribution of benefits could undermine ethnic community support for tourism, consequently reducing the potential for sustainable development.

Differences were identified in marketing ethnic cultures. The Chinese case focused on marketing the contemporary ethnic minority culture, while the Canadian case put greater emphasis on the historical perspective of the aboriginal culture. In the YFCV, a selective procedure was used to decide the culture programmes – only those judged as reflecting 'healthy and progressive aspects of minority life' were presented. By doing so, the authenticity of ethnic culture could be compromised, in favour of showing merely the unity of the Chinese nationalities under the leadership of the Chinese Communist Party (China National Tourism Administration [CNTA], 1994). In WHP, the focus on pre-contact aboriginal culture possibly catered to some tourists' need to seek a primitive and exotic 'other' people who are different from the Caucasian tourists' modern life. Through this, WHP might be reluctant to recognise contemporary aboriginal culture, thus inadvertently creating a false impression that the aboriginal culture would never change.

Ethnic responses and adaptations to tourism development also manifested differently between the two case studies. In the Chinese situation, the perception of tourism as a 'gold rush' seemed to have a substantial impact on the operation of the YFCV. Commercial activities appeared to be the dominant theme with all of the ethnic culture presentations. For instance, there were many souvenir arts and crafts shops and live ethnic dances were performed throughout the day for paying customers. Such commercial activities lead to ethnic people adapting to modern commercialised society, thereby eroding their traditional culture and social values, which are vital for the survival of ethnic tourism. In contrast, ethnic involvement in the economic development appeared to be limited in WHP. Although healing the gap between the aboriginal people and other Canadians was precisely what WHP tried to accomplish, such a healing process would require supportive social and economic environments. Since many socioeconomic disparities still exist between the aboriginal and non-aboriginal communities, ethnocentric distrust and prejudice could make cooperation between both very difficult to achieve. Without the cooperative efforts of both the ethnic groups and mainstream societies, the economic sustainability of WHP would be out of the question.

These challenges to coping with inherent issues in tourism development also manifested differently between the two cases. In the Chinese case, a key barrier to ethnic control appeared to be the nation-state labour policy that restricted rural residents from securing long-term stable employment in cities. Since ethnic minority people were mostly rural residents at the time of this research, they might only be hired on short-term contracts that were largely operational, non-managerial positions. In the Canadian case, the roots of European dominance over the institutionalised Canadian culture system appeared to be a stumbling block to aboriginal people's participation in planning and managing tourism businesses, although Canada had various politically right policies catering for aboriginal peoples. An aboriginal storyteller pointed out:

> I don't have a university degree, and I don't have what they call a 'formal education'... but I learned my culture from my elders, and I have education which came to me from a different system... but they [the non-aboriginal group] don't think so. (Author's direct interview quotation)

Both cases faced the dilemma of ethnic culture freezing and evolution – another major challenge to ethnic tourism development. In the Chinese case, fast social, cultural and economic changes accompanying ethnic tourism development overwhelmed the attempts to conserve ethnic traditions and customs. In contrast, the Canadian case manifested a different scenario. The tendency to present a pre-contact culture might have neglected the social, cultural and economic progress of the aboriginal community. This progress is crucial to their self-identification as part of Canadian contemporary society. Neither of the above-mentioned tendencies – fast change without appropriate conservation and ethnic culture freezing – can be considered conducive to achieving sustainable ethnic tourism development.

The strategies adopted for reducing those challenges were found to be different between the YFCV and WHP. Comprehensive and sophisticated tourism impact assessments were completed in the Canadian case (WHP, 1992a). In the Chinese case, the YFCV did have certain strategies for addressing development issues; however, they only focused on the economic aspect. Little attention was directed to the cultural and environmental aspects. One of the managers claimed that the YFCV did have marketing and training plans to resolve development issues. This claim, however, probably reflected the reluctance on the part of the YFCV to publicise its development agenda. On the other hand, although WHP showed a comparatively more advanced stage of development with its prepared impact assessment reports, resource management and marketing and training plans, a thorough examination of those plans revealed certain problems. For instance, the impact assessment merely focused on the conservation of cultural and natural environments, with no measures for adapting and adjusting to the possible social/cultural and economic changes to the aboriginal community. WHP had comprehensive strategies regarding the use of the traditional aboriginal culture but neglected the importance of developing contemporary culture programmes to present a complete image of aboriginal people's contemporary life.

Positive achievements

Notwithstanding those challenges, positive achievements were made in both cases. Those achievements are categorised into three realms in line with Hinch and Li (1994; see Figure 2.2). For the social/cultural realm, both practices of the YFCV and WHP indicated possible shifts towards increasing ethnic control. In the Chinese case, this shift was a result of the nation-state policies of political and economic reform that encouraged the ethnic minority communities to invest in tourism ventures. As long as this official support for the commercial economy continued, ethnic control of tourism operations would materialise in the process of interactions among government agencies at the national and local level, international private tour companies and, most importantly, local ethnic minority entrepreneurs. As for WHP, the hope for a potential increase in ethnic control was grounded on the belief that a successful tourism business would help build an economic base conducive to ethnic involvement. An eventual increase in economic independence would lead to ethnic cultural independence.

The tourism practices in both cases appeared to function as accelerators for positive social/cultural changes to ethnic life. In the Chinese case, the increasing number of ethnic minority people participating in the tourism business liberated them from traditional farming life. A prominent example was that ethnic women's social status was enhanced with their participation in the tourism business, according to a female tour manager of ethnic minority background:

> Look at these street vendors selling arts and crafts here, they are all women. The booming tourism business is changing the gender role in our community. It helps to raise women's social status. Don't you think it a positive sign of social change in our minority society? Women were traditionally less valued than men in our communities. (Author's direct interview quotation)

In WHP, the training programmes included a wide range of knowledge about Canadian aboriginal culture, from the perspectives of archaeology and natural and cultural history, as well as contemporary administrative issues (WHP, 1992b). This improved the understanding of the merits of traditional aboriginal knowledge, thus benefiting both aboriginal and non-aboriginal peoples. As stated by a government official of aboriginal background:

> All that [contents of the training plan] is very educational... especially when they get a lot of school groups... The school kids have the chance to learn the way that their ancestors survived in pre-contact times... Time and technology have changed, but it is still important for kids to understand that the technology our people used hundreds of years ago is still right for protecting the environment. (Author's direct interview quotation)

Both cases were eager to develop and protect ethnic culture. The Chinese ethnic minority groups had been isolated from mainstream culture for generations. Tourism development opened a window on cultural exchange, thereby enhancing national and international understanding of ethnic minority cultures. The cultural exchange and understanding processes offered visitors the opportunity to appreciate ethnic minority cultures and to develop an ethnic sense of value in support of cultural conservation. In WHP, the positive presentation of aboriginal history and cultural values improved the way that non-aboriginal people view aboriginal culture and history. Canada's past practice of detaching aboriginal people from their heritage blinded many non-aboriginal Canadians to the merits of aboriginal culture. WHP is considered a first step in the education of society about aboriginal cultures in Canada. As this two-way education introduces new ideas to the surrounding aboriginal communities, tourism functions as a means of reinforcing the positive aspects of the uniqueness of ethnic culture to both aboriginal and non-aboriginal peoples. This education process informs the ethnic minority groups that it is the essence of their culture that identifies ethnic people from tourists. In this way, WHP's presentation of aboriginal culture could change the stereotypes that some Euro-Canadians have of aboriginal people, thereby contributing to a broader awareness of the need to develop and protect aboriginal culture in Canada.

This research found that both cases had been successful in creating jobs for ethnic people, supporting ethnic-owned businesses and offering potential for further economic development within ethnic communities. The YFCV not only created jobs for many people of ethnic minority groups, but also functioned as a training centre for qualified ethnic tourism professionals. Additionally, the marketing of ethnic arts and crafts benefited private businesses owned by ethnic minority groups because most of the handicraft products were 'cottage crafts' made by ethnic minority households. In the Canadian case, the fact that 90% of the workers were of aboriginal background (Morgan, 1994) indicated tourism's considerable contribution to increasing aboriginal employment. In addition, the opportunities that WHP created for aboriginal people to run a business, such as the gift shop and the restaurant, enhanced the ethnic sense of independent entrepreneurship. Such entrepreneurial role models stimulated the growth of other ethnic-owned businesses, contributing to the creation of a strong economic base conducive to ethnic involvement in tourism planning and management.

In terms of environmental protection and conservation, positive signs were identified in both cases. To a large extent, the YFCV served as a 'honeypot' that provided a 'staged authenticity' (Fuller *et al.*, 2022; Ryan, 1991; Tang & Liang, 2022) for tourists to experience ethnic cultures. This 'honeypot' function helped keep mass tourism development within purpose-built ethnic cultural sites instead of introducing tourists to ethnic minority hinterland areas, thereby relieving the tourism pressure on naturally and culturally sensitive ethnic places. In the Canadian case, environmental education appeared a major aspect of the cultural programme. An important factor that contributed to the establishment of WHP was the perceived need to revert to tribal roots and the wisdom of elders to solve global ecological problems (Kylie, 2017). WHP informs visitors about the traditional aboriginal lifestyle and its spiritual and environmental values. Such environmental education is necessary to the contemporary world where ecosystems are endangered by various irresponsible human actions.

Conclusion

The research findings suggest that ethnic tourism development is not without problems due to various social/cultural, economic and environmental issues inherent in tourism development. However, it is possible to reduce those issues, considering the evidence of positive achievements on both cases. A basic requirement of any tourism development is to increase positive achievements and reduce challenging issues. Accordingly, future ethnic tourism could consider the following recommended directions.

First, the Western discourse on non-Western cultures as the so-called 'exotic' and 'other' ones must be deconstructed along with the globalisation of the world's peoples and places. In line with Western discourse on the 'Orient', contemporary tourism contexts have constructed various images of non-Western people and places as exotic and quaint 'others' – cultural images that often appear different from 'civilised' Western people and places (Shim *et al.*, 2015). Implicitly and explicitly, the ethnic tourism literature portrays those 'others' as precious, rare or endangered 'species' in contrast with 'mainstream' Western ideas, achievements, values and beliefs. Thus, ethnic people and places need to be protected and conserved against the current waves of globalisation – a euphemism that suggests anything ethnic should remain static against change. However, tourism itself is a process of change in the social/cultural and economic evolution, especially the contemporary situation in China, which has changed at unprecedent speed. The prevalent Western discourse of 'Orientalism' may find tremendous difficulty in the application of China-related studies. The focal point of view 'Orientalism' is grounded on the imperialist and colonialist legacies of the Near East and Palestine, leaving scant attention for China and East Asia. The sustainable development of world tourism in general, and ethnic tourism in particular, should not overlook this important part of world culture and heritage. Otherwise, ethnic tourism development would not be able to achieve the delicate balance between conservation and change, which is crucial to the sustainability of any type of cultural tourism (Li, 2003a).

Second, barriers to the ethnic control of tourism practices must be reduced throughout the development process, building a solid economic base for ethnic involvement which is key. The government policy of both China and Canada is ethnic autonomy (Chen, 2008; Hoerder, 2010). However, related policies appear difficult to fully exercise because of certain social/cultural and economic constraints. In the Chinese case, it is essential that developers find a way through the national human resources' regulations to enable more qualified ethnic minority people assume managerial positions. As the Chinese central government's present policy of reform encourages more input from local decision makers, the YFCV should take bold and creative actions to develop the ethnic tourism site, instead of passively adjusting to national policy change. In the Canadian case, the function of education should be strengthened. The barriers to exercising ethnic control are not only economic, but also social/historical. It is important that WHP continues to help the aboriginal people build a sense of identity and self-esteem. WHP bears the responsibility to encourage non-aboriginal societies to appreciate ethnic culture, thereby creating space in mainstream society for ethnic culture to exist. It should also seek alternative measures to diversify the operational budget sources, in order to minimise financial dependence on outside government agencies. Otherwise, the independent power of WHP would not be strong enough to exert a larger social influence. Cultural commodification is inevitable for any type of ethnic tourism. In both cases, this commodification process must eventually facilitate the exchange of ethnic culture for financial return to the ethnic groups involved with the process, thereby establishing a relevant economic base to strengthen ethnic autonomy.

Third, the inherent social/cultural issues of ethnic tourism must be recognised and addressed, to encourage development opportunities and reduce challenges. For the YFCV, economic drive should be balanced with the conservation of cultural and natural resources that are vital to long-term development. The tendency to use ethnic tourism merely for economic benefits must be avoided and grounds for the sustainability of ethnic culture must be established in the development process. In doing so, the commodification of an ethnic culture for tourism must be sufficiently controlled by the ethnic groups. That ethnic minority people are located on the periphery of the decision-making process must change. As for WHP, the tendency to freeze ethnic culture must be altered to permit positive change, so that a complete aboriginal image, including the living and vibrant culture, can be presented. One of the educational objectives of WHP is to change the stereotypical image of the aboriginal population held by the dominant society in the past. WHP should look at the aboriginal culture from both a historical and a contemporary perspective, balancing the ideas of conservation and change in its mandate to protect and develop aboriginal cultural heritage. A frozen ethnic culture, with no presentation of its contemporary life, would serve to reinforce the stereotypical image of the aboriginal population as a primitive people. If this stereotypical image were produced by inappropriate tourism, it would affect aboriginal people's sense of identity and self-worth, even while providing economic income. For both the YFCV and WHP, the environment vital to the success of ethnic tourism consists of both cultural and natural aspects, neither of which is static. To maintain a healthy cultural environment for successful development, conservation is not enough because, historically, cultural change has taken place accidentally, such as new ideas or techniques being transmitted from one culture to another without any deliberate effort (Arensberg & Niehoff, 1971; Whelan, 2016). Accordingly, both the YFCV and WHP should develop relevant social, cultural and economic strategies conducive to ethnic tourism development.

To conclude, development strategies must seek to achieve an appropriate balance between conservation and change in the pursuit of sustainable ethnic tourism. A destination changes and evolves over time, with the paradox that tourism can impair and even destroy the very qualities of the environment that attracted tourists in the first place (Edwards, 1988; Ferrara, 2017). Theoretically, built tourism attractions are planned and managed to avoid this consequence (Gunn, 2002). However, in practice, both the YFCV and WHP still have to adopt appropriate strategies to address the paradoxes that tourism may lead to mis-results.

In sum, the operation and management of both case studies appeared consistent with the economic perspective of ethnic tourism (see Figure 2.1). Ethnic culture is a resource, but the tourism products are the interpretative programmes offered at both the YFCV and WHP. This research demonstrated that benefits and costs coexist, respectively, in both cases, with regard to the three realms (see Figure 2.2). To pursue maximum benefits and minimum costs for sustainable development, future ethnic tourism practitioners need to consider an important direction. Namely, the improvement of sustainable ethnic tourism rests on a three-pointed foundation: theoretical and practical confirmation of ethnic guidance and leadership; overall identification of tourism's inherent social/cultural and economic issues; and appropriate strategies for resolving the issues.

3 Tourism Impacts in China after Two Decades of Development

Introduction

The development of a tourism destination can have social/cultural, economic and environmental effects upon a local community, which are defined as tourism impacts (Comerio & Strozzi, 2018; Williams, 1998). Such impacts manifest either as benefits for local communities or as barriers to the communities' acceptance of tourism development (Nickerson, 1996). By the turn of the century, tourism researchers had studied the impact of tourism on local societies in many parts of the world (Bodley, 1990; Boniface & Fowler, 1993; Cohen, 1996; MacCannell, 1992; Mathieson & Wall, 1997; Poirier, 1995; Wilson, 1996), but few had examined the impacts on Chinese local communities. A few studies considered the impact of tourism on China's ethnic minority groups (Adams, 1996; Li & Hinch, 1997; Oakes, 1995, 1998; Swain, 1989, 1990, 1993, 1995; Toops, 1995). However, those studies only focused on the geographically remote and peripheral regions of China, thereby examining tourism impacts on only a small portion of the Chinese population. Although China's official media frequently publicises the positive aspects of the tourism industry, as always, it only presents the 'good' aspects of tourism due to the nation-state dominance of national discourse. This chapter presents a study that was conducted in the early days of the 21st century on the positive and negative effects of tourism on a Chinese local community. It aims to provide useful insights for scholars and researchers who are interested in the study of growth and change in contemporary China.

The chapter begins with a brief literature review of the nation's tourism policy documents since 1978, in order to inform readers about the evolution of China's tourism development over two decades (1980–2000). It then focuses on some specific tourist activities in China during that period, such as ethnic tourism, in order to contribute to the knowledge needed to minimise the costs and maximise the benefits of tourism. A case study approach is applied to investigate cross-border tourist activities in Shenzhen – a border city neighbouring Hong Kong – in order to analyse some specific tourism impacts on a Chinese local community during that period. The social/cultural, economic and environmental aspects associated with tourism development are considered in the analyses. A conceptual model is proposed for studying encounters between tourists and hosts at tourism destinations, where maintaining a sustainable social/cultural, economic and natural environment is imperative for the survival and progress of the local community.

Literature Review

A historical overview of tourism development in China

Leisure travel in China dates back thousands of years. Emperors, scholars and monks were frequent travellers in ancient China, whether for leisure or political or intellectual interests. However, travel to China as an activity of modern consumption only started when Thomas Cook and Sons, among other world-renowned travel companies, opened their offices in the country during the 1920s (Hibbort, 1990). For a period of time, wealthy and adventurous foreigners controlled international travel to China and dominated the travel industry as both participants and business operators. Thomas Cook and Sons (China) was initially located in Shanghai, later moving to the old Hotel de Pekin in the country's capital. The office handled ground services for international and domestic tours in China. Between the 1930s and 1940s and the war years, all of China's leisure travel came to a standstill. After the People's Republic of China was founded in 1949, the communist state utilised the travel business as a strategic political tool. Travel services (agents and operators) were set up to provide services only for inbound overseas Chinese visitors and foreigners with special permission to visit China. Hence, for some time, tourism in China was essentially a 'diplomatic mission' and was more politically than economically operated (Zhang, 1995; Zhao & Liu, 2020). Domestic tourism activities were virtually non-existent, and outbound travel was limited almost exclusively to diplomats and government officials. The chaotic 'Cultural Revolution' movement from 1966 to 1976 once again delayed the development of China's tourism industry. It was not until 1978, when the late paramount leader Deng Xiaoping initiated China's reform and opening up, that tourism started to be considered both as a gesture of embracing the world and as a vehicle for driving the economic reform. In the course of those reforms and China's increasing openness, nation-state policies moved towards economic development. As one component of the nation's growing economy, the tourism industry has also seen the government's policies change in line with the general emphasis on the nation's economic development.

The official purpose of tourism in China went through three major phases: pure politics, politics plus economics, and economics over politics (Zhang, 1995). The first phase covers the early period of communist rule from 1949 to 1978. During that period, tourism in China was nothing more than a component of the country's foreign affairs policy. Focusing on so-called 'people-to-people diplomacy' (Zhu, 2001), the tourism sector sought no economic benefits for the country. Tourists were either inbound overseas Chinese visitors or 'foreign friends and guests', and tourism was a means for the communist nation-state to cultivate international friendship, understanding and sympathy from the world. Overseas tourists, although small in number, were treated as VIPs at endless banquets, meetings with leaders, courtesy calls and visits to work units. The Chinese hosts arranged the travel schedule regardless of the real interests of the guests. The general tourism practice in China during this phase was that the destination (China) selected the tourists rather than the tourists choosing the destination.

The second phase was during the early stages of China's economic reform (1978–1985), when the country started to consider tourism as both a component of foreign affairs and an economic activity. However, the central government's tourism development policies still placed politics above economic gain. One example was a discriminatory pricing policy that allowed overseas Chinese and compatriot residents from Hong Kong, Macao and Taiwan to pay considerably less than non-ethnic Chinese international visitors for the same consumer activities. Prices and services varied according to a tourist's race and ethnicity rather than his/her willingness to pay. Occasionally, overseas Chinese or compatriots were refused by hotels or transport ticket offices, or they received poor service, because they paid less than non-ethnic Chinese international tourists. International tourists, on the other hand, also complained of being embarrassed and annoyed by their preferential treatment (Wang *et al.*, 1997).

From 1986, China's tourism industry entered its third phase, when the government declared tourism to be a comprehensive economic activity with the direct purpose of earning foreign exchange for the nation's modernisation campaign. For the first time, tourism was included in China's national plan for social and economic development. Since 1986, the central government has repeatedly stressed the importance of tourism as an essential service industry that requires less investment, but produces faster profits, larger employment potential and greater opportunity to improve people's livelihood than many other tertiary service sectors (Zhang, 1990). Central government made the development of tourism one of its key industrial policies, placing its economic benefits foremost, in contrast to the political emphasis of past years. Even through the recession that resulted from the 1989 Tiananmen events, in 1992 the central government decided to host China's first themed tourism year in its history – 'Visit China '92' – which was considered an epoch-making moment in the country's tourism industry.

Emerging impacts of tourism

The three phases of tourism development brought China tremendous economic benefits around the turn of the century. The net receipts from tourism accounted for approximately one-third of the country's foreign exchange earnings from its non-commodity foreign trade (Tisdell & Wen, 1991; Zhao & Liu, 2020). Foreign and domestic tourist arrivals reached 38 million in 1992, representing a 14.3% increase on the previous year. Four million of the arrivals were foreigners, reflecting a 47% increase on the previous year (China National Tourism Administration [CNTA], 1993). China sustained this growth, becoming a leading tourism destination (Wayne, 1999). In 1999, China's tourist arrival figures were ranked top five and tourism receipts top seven in the world (CNTA, 1999). That year, China's tourist arrival figures reached a record high of 72.796 million (CNTA 1999), with the tourism industry contributing US$87 billion to China's gross domestic product (GDP) and providing jobs for 49.4 million people, accounting for 7% of China's total employment (Wayne, 1999). On 13 July 2001, the International Olympic Committee selected Beijing as the host city for the 2008 Summer Games (Hersh, 2001). This created a new incentive for the Chinese government to focus on the economic benefits of tourism in a bid to accelerate the country's rapidly growing economy.

While China benefitted greatly from tourism growth in economic terms, the impacts of tourism also transformed the nation's social/cultural, economic and natural landscapes, particularly at the local level. Scholars from home and abroad raised concerns about the commoditisation of ethnic minority cultures through so-called 'ethnic tourism' (Hinch & Li, 1994; Li & Hinch, 1997; Oakes, 1995, 1998; Swain, 1989, 1990, 1993, 1995). Ethnic tourism is an activity that involves some form of direct experience with an ethnic culture and its environment, often with reference to the 'quaint' customs of indigenous and other 'exotic' peoples (Smith, 1989). However, some Chinese ethnic minority cultures attach enormous symbolic and spiritual importance to their ceremonies and art objects (Ma, 1999), so an adequate interpretation of their symbolic and spiritual significance requires considerable anthropological knowledge on the part of the consumer (Oakes, 1995, 1998).

Traditional Tibetan funerals, for instance, were repeatedly described as a tourist attraction in Chinese travel magazines and brochures (Adams, 1996). Without an understanding of the sensitivity and cultural significance of this tradition, tourists would merely see a Tibetan funeral as an 'exotic' or 'quaint' custom, leading to misunderstanding and conflict between tourists and the local people (Hinch & Li, 1994). Similarly, large crowds of affluent tourists in tourist regions could lead to local resentment if there was a substantial disparity between the demonstrated wealth of the tourists and that of their ethnic hosts (Yang *et al.*, 1999). Tourists form a reference group for the visited community by providing evidence of the relative affluence of other nations, leading to local resentment towards tourists (Bryden, 1973; Yasothornsrikul & Bowen, 2015). Supporting this point is an incident that took place at the Lake of Thousand Isles in Zhejiang Province. On 31 March 1994, the passengers and crew of a tour boat died in a fire set by three local youths after they had robbed the affluent tourists (Sun, 1994). This tragedy might be an extreme case, but it demonstrated the tourism impacts of large disparities in wealth.

Around the turn of the century, the ever-increasing impacts of tourism activities on the ethnic minority communities in China drew the attention of international scholars from various fields. Well known among them is Margaret Byrne Swain, an anthropologist from the University of California at Davis. She had studied the tourism impacts on China's southwest Yunnan Province for many years, and attempted to analyse a number of ethnic tourism issues including majority Han Chinese dominance versus the cultural hybridity of ethnic cultural regions, the negotiation of ethnic identity in tourism interpretation, as well as gender and class relations in ethnic tourism business operations (Swain, 1989, 1990, 1993, 1995). Other scholars, such as Li and Hinch (1997), Oakes (1995, 1998), Toops (1995) and Adams (1996), also studied the social/cultural, economic and environmental impacts of ethnic tourism activities on the minority communities. However, they focused more on the remote ethnic minority regions and considered the impacts on a very small portion of the Chinese population. Those studies were far from comprehensive in terms of informing about the contemporary growth and change of China under the impact of tourism development. The case study of this chapter represents a timely intervention during that period of time, into the impacts of tourism on a tourist destination inhabited by the majority Han Chinese people. The focus of this case was Hong Kong tourists' cross-border holiday travel to Shenzhen – China's first special economic zone (SEZ) and the 'window to the outside world'.

Research Methods

Research case selection

The research case – Shenzhen – is located to the north of Hong Kong along the southern coast of China and is China's first SEZ. By the turn of the century, Shenzhen had developed into a metropolis of 2050 km^2 with a population of 3 million (Shenzhen Statistics Bureau, 1999; Yu *et al.*, 2019). The Chinese central government designated Shenzhen an SEZ in August 1980, which accelerated the city's economic growth including its tourism development. In approximately two decades, the city's fixed assets investment in tourism had reached US$12 billion (Shenzhen Statistics Bureau, 1999), and its tourism income had reached RMB200 billion (US$1 = RMB8.3) in 1999, accounting for 16% of its GDP for that year (Ming Pao Daily, 2000). By August 2000, the city's tourist arrivals had reached 13 million, a 1.5% increase on the same period in the previous year, and its tourism income had reached RMB160 billion. The tourism industry became a mainstay of the city's economy and brought various economic benefits. However, there were also negative effects such as inflation, prostitution and crime (Ming Pao Daily, 2000).

Table 3.1 Room rates of selected hotels in Hong Kong and Shenzhen

Hotels in Hong Kong	Star	Rate (HK$)	Hotels in Shenzhen	Star	Rate (HK$)
Mandarin Oriental Hotel	5	1,400	Shangri-La Hotel	5	765
Hotel Nikko	5	1,180	Panglin Hotel	5	660
Grand Stanford Inter-Con.	5	1,080	Sunshine Hotel	5	610
The Harbor Plaza Hotel	5	780	Nan Hai Hotel	5	590
Sheraton Hotel	5	760	The Landmark Hotel	5	580
Hong Kong Hotel	5	730	Forum Hotel	5	480
The Excelsior Hotel	4	1,750	Century Plaza Hotel	4	460
Renaissance Harbor View	4	830	Shenzhen Bay Hotel	4	418
Harbor Plaza North Point	4	820	Hotel Oriental Regent	4	410
Newton Hotel Kowloon	3	630	Guang Dong Hotel	3	400
Grand Tower Hotel	3	560	Bamboo Garden Hotel	3	330
Park Hotel	3	560	Xili Lake Resort	3	320
Newton Hotel	3	530	Oct Seaview Hotel	3	280
Stanford Hillview Hotel	3	520	Honey Lake Country Cl.	3	260
YMCA	3	520	Cits Hotel	3	240
BP International House	3	480	Siu Mui Sha Resort	3	238

Adapted from China Travel Service HK Ltd. (2000)

Of the contributors to the rapid growth of Shenzhen's tourism profits, cross-border holidaymakers from Hong Kong were probably the most prominent. Hong Kongers crossed the border and travelled north to enjoy shopping and other leisure activities at lower retail prices. One example of the more attractive prices was the comparison of hotel room rates between the two cities (Table 3.1). Traditional definitions of tourism often exclude such cross-border leisure activity seekers, who are not regarded as 'real' tourists because they usually stay in a destination for less than 24 hours. However, some tourism researchers (Timothy & Butler, 1995) argue that the definition of tourism should include such short-term travel, because it involves crossing an international border and frequently has pleasure as the purpose of the trip. It appears in the Hong Kong/Shenzhen context at least, and probably also in Europe and North America, that cross-border shopping and other leisure pursuits are powerful motivating factors when making the decision to take a holiday. Thus, the cross-border leisure activities of Hong Kong weekend holidaymakers to Shenzhen are considered the best fit for a case study of the tourism impacts on a Chinese local community during that particular period of time.

Investigation techniques

This specific investigation of the tourism impacts on Shenzhen was carried out from 2000 to 2001, based on data collected via an in-depth interview survey, documentary research and some indirect sources, such as newspapers and other media reports, which were initially collected for other purposes. The interview survey was verbally administered in the form of two open-ended questions to allow for qualitative analysis. The questions concerned, in very general terms, tourism's positive effects and negative impacts on the local community, thus allowing the respondents to cite relevant instances of those effects and impacts. One hundred respondents were selected through availability sampling, a typical form of which is to interview people in a shopping mall to obtain a sample of the community (Andrade, 2020; Marlow, 1998; Oliveira *et al.*, 1992). For this investigation, the East Gate Market in the city centre of Shenzhen was chosen as the sampling location. This is a popular leisure shopping area for both local residents and Hong Kong cross-border holidaymakers at the weekend, and was considered appropriate for achieving a sample of local residents who had lived in the city for more than two years and had encounters with Hong Kong tourists at the market. Of the 100 residents approached, 15 refused to participate in the survey (a refusal rate of 15%).

Research Findings

The results of the research suggested that the respondents perceived a multitude of tourism impacts on their community, which were manifested, specifically, as follows. The first was that the respondents saw the benefits of tourism mainly in economic or economy-related social/cultural terms, including its capacity to attract foreign capital investment, create job opportunities, improve living standards and enhance community life by offering better consumer goods and services. This indicated that tourism had served as a means of economic growth for Shenzhen. According to Chen and Yu (1992), Shenzhen's tourism development has gone through three phases since 1980: the business, culture and future phases. The first – from 1980 to 1988 – was the business travel phase, when Shenzhen became attractive to outside investors, particularly those from nearby Hong Kong, after the central government had designated the city as an SEZ. Several country clubs, such as the Honey Lake Country Club and Xili Lake Resort, were built to cater to business travellers. The second phase – from 1989 to 1999 – was the culture travel phase, when a number of cultural theme parks were built, mainly targeting cross-border weekend day-trippers from Hong Kong and the emerging Chinese domestic tourist market. The most prominent cultural theme parks built during this period include Splendid China, Window of the World and China Folk Culture Village. At the start of the 21st century, the Shenzhen municipal government implemented the third phase of tourism development – the so-called 'future phase', to fulfil the overall economic growth target of the new century.

Table 3.2 Shenzhen foreign exchange earnings from tourism

Time	Total foreign exchange earnings of tourism (US$10,000)	From commodities (US$10,000)	From services (US$10,000)	Room occupancy of hotels and resorts (%)
1990	20,933	11,389	9,544	68.1
1991	20,911	6,410	11,501	72.3
1992	39,100	14,774	24,326	79.6
1993	39,128	15,068	24,060	68.0
1994	48,821	17,483	31,338	63.0
1995	63,633	20,574	43,059	58.9
1996	58,381	21,489	36,859	56.3
1997	53,225	17,176	36,049	57.6

Adapted from Shenzhen Statistics Bureau (2000)

As one of the components of Shenzhen's economic growth, tourism helped increase commodity output, create jobs and, most importantly, bring wealth to the city (Table 3.2). The city's total foreign exchange earnings from tourism in 1995 were more than double those of 1990. Increased tourist expenditure not only contributed to the local economy but also generated income multiplier effects, giving rise to a chain of expenditure–income–expenditure until leakage occurred (Goeldner & Brent Ritchie, 2011). Increased tourism expenditure would also stimulate business ventures, and more business ventures would create more jobs. This income multiplier effect was evident in Shenzhen's economic growth. By 1999, tourism-related businesses had experienced an average annual growth rate of 31.2% for two decades (Shenzhen Statistics Bureau, 1999). The increase in business led to an increase in employment and income, and ultimately an increase in local purchasing power. Within 10 years (1989–1999), the average annual rate of income growth of Shenzhen's employed residents was recorded at 18.1%, and retail sales in consumer goods grew at an average annual rate of 35.2% (Shenzhen Statistics Bureau, 1999).

The second point made by some of the survey respondents was that tourism had functioned as a process of social/cultural and economic transformation for the city. The inflow of outside visitors not only brought economic benefits but also made tremendous social and cultural change. They voiced their concerns about the increase in the problems associated with those changes, such as social instability, prostitution, family crises and conflicts between local residents and outsiders. They believed that the rapid economic growth and the social and cultural change in Shenzhen had attracted not only tourists but also large and increasing numbers of migrant workers who dreamed of making a better living in the SEZ. A few respondents were worried that some migrant workers might become beggars, or even turn to crime when their dreams were not realised, consequently leading to greater violence and crime in the SEZ. Within 20 years, Shenzhen had developed from a small farming village into a modern metropolis with more than 3 million residents. The local society became more complicated with this rapid inflow of people, and the increasing crime rate was not the only price that Shenzhen paid for its rapid economic growth and social change. Prostitution, particularly catering to tourists and business travellers to the city, became quite common, although this had rarely been reported 20 years before. One respondent described the situation:

> Walking fifty metres down the road in front of the Century Plaza Hotel near the Luo Hu Railway Station, a man probably would be approached more than ten times by greetings such as: 'Do you want girls?' 'Would you like a massage? We have the best girls in town'. Prostitutes are everywhere in Shenzhen. They are so obvious, and they pollute the spirit of the society, making Shenzhen notorious for 'yellow business'.[1] (Author's direct interview quotation)

A consequence of the booming 'yellow' business was probably the rapid increase in venereal disease. A Hong Kong news media reported that Shenzhen had at least 78 AIDS patients, and there were at least 2000 people with the HIV virus in the region (Oriental News, 2000). As Hong Kong residents enjoyed a higher income, Shenzhen became increasingly popular among Hong Kong residents as a weekend cross-border leisure retreat. From January to June 2000, total retail sales in Shenzhen reached RMB250 billion, one-third of which was attributed to Hong Kong cross-border holidaymakers (Apple Daily, 2000). This increased local inflation, and consequently inflamed local resentment towards wealthier visitors from Hong Kong, as revealed by a local real estate agent:

> I am doing approximately 70 percent of my business for Hong Kong home owners. They bought flats in Shenzhen because it is cheaper here. They come to buy flats, electronics, furniture, everything you can imagine. They even buy our girls. From time to time I see ugly old guy from Hong Kong make housing arrangements for their local secret second ladies (concubines). They don't care about the human dignity of the women. The worst thing is, renting concubines has become a trend here. It is immoral and causing family crises. It hurts me. It is really horrible. (Author's direct interview quotation)

The third point made by the survey respondents was that tourism development had also manifested a process of environmental modification of the city, although they perceived the impacts of tourism to be mainly from the social/cultural and economic aspects. They saw the environmental benefits of tourism as leading to the greening of the city, preserving or revitalising heritage sites and promoting environmental awareness. Even though the respondents provided little information on the perceived negative environmental effects of tourism, they did identify a serious issue that needs to be resolved for both Shenzhen and China as a whole. Tourism development has transformed the natural environment through modification of urban landscapes and the destruction of wildlife habitats. Some respondents criticised the threat posed by Shenzhen's cultural theme parks to the Mangrove Tree Natural Reserve nearby, pointing to a problem in the city's tourism planning strategies, particularly in terms of the construction of theme parks.

The three theme parks – Splendid China, China Folk Culture Village and Window of the World – were built right beside Shenzhen Bay. Some of the land used for building the parks was reclaimed from the ocean. The success of the parks thus encouraged Shenzhen's urban planners to reclaim more land from the ocean in order to develop the city into a garden-like beach resort (People's Daily, 2000). As part of the city's master development plan, a brand-new highway, Bin Hai Da Dao (Harbour Front Avenue), was constructed on reclaimed land, running through the three parks, posing further threats to the natural habitat of the wildlife, particularly the black-faced spoonbills in the nearby Mangrove Tree Nature Reserve. The construction of both the parks and the highway required modification of the natural environment, and was considered a disaster by some respondents. However, the parks themselves are popular with tourists from both Hong Kong and other parts of China (Chen & Yu, 1992).

Splendid China presents the cultural and natural wonders of China through miniature reproductions, such as the Great Wall, in order to provide a glimpse of China's cultural and natural landscapes. China Folk Culture Village showcases the folk customs and traditions of China's ethnic minorities, with reproductions of life-size ethnic villages and live performances by ethnic minority artists, as a taste of China's rich ethnic cultures. Window of the World has 118 replicas of well-known historical, cultural and scenic spots from around the world, such as the Sydney Opera House, providing visitors with a quick guide to the world. The success of these theme parks was largely due to their popularity among Chinese domestic tourists, and this success had implications for some strategic issues related to tourism development. Most Chinese people's knowledge about nature and the natural landscape was gained from literary classics including poetry and paintings (Cui *et al.*, 2017; Peterson, 1995). Many Chinese people were familiar with famous places and landscapes before they actually saw them on-site and tended to appreciate places and landscapes through illustrated images rather than the harsh physical experience of nature and the wilderness. The popularity of these cultural theme parks might be attributed to the fact that they only cater to this traditional taste, by offering a mirror of the classical places portrayed in Chinese literature and the arts. Their future success, however, was in doubt, with the continuous exposure of the Chinese people to global cultures and norms.

Conclusion

At the beginning of the 21st century, after two decades of opening up the country to the outside world, China's culture and society had transformed. This led to significant changes in the economic structure, the social and cultural values of the Chinese people and the ideological thinking of the political leaders. As an essential part of China's attempts to embrace the world economy, the tourism industry accelerated the processes of change, which had various social/cultural, economic and environmental impacts on local communities. There were both positive and negative aspects to the impact of tourism, consistent with the findings of similar studies of other societies of that time (Boniface & Fowler, 1993; Cohen, 1996; MacCannell, 1992; Mathieson & Wall, 1997; Wilson, 1996). The discussed impacts of tourism on a Chinese local community indicates that China's tourism growth could inflict substantial negative effects upon a people even though the impact also created impressive economic benefits. Economic growth is essential to China's agenda to modernise the nation (Zhao, 1987, 2021); however, the negative effects of development must be noted and minimised. Otherwise, the people may not support the expansion of tourism, and the issues of social and economic transformation associated with tourism growth may not be appropriately resolved.

Tourism is a system that involves a sum of relations at the destination, as discussed in Chapter 2 (also see Figure 2.1). The market is composed of those people who seek a unique experience at a destination. This market demands certain resources to facilitate that experience. However, like many other types of economic activity, further processing of basic resources is required. While the raw resources at a destination may provide the context and environment for tourism, further refinement is necessary to produce a tangible product that is appropriate for tourist consumption. The tourism literature commonly refers to this process as the commodification of resources (Greenwood, 1989). Nonetheless, negative connotations are often attached to this process based on the argument that commodification has negative impacts on a destination, such as the erosion of the culture's authenticity.

It can be countered, however, that commodification may actually represent a mechanism for protecting the resource. Destinations that selectively transform their resources into tangible products not only facilitate the exchange of the tourist experience for a fair financial return, but also create a situation in which they can promote sustainable development through the careful management of their resources (Craik, 1991). While the market's original demands are focused on the resources, the actual consumption will be of a processed product. The physical travel of tourists occurs between the location of the market (origin) and the location of the product (destination). To the extent that the production and distribution of this product are controlled by the destination (temporal and spatial), the host communities have the opportunity to protect their resources of both cultural and natural significance.

The 'tourism system of resource commodification' (Hinch & Li, 1994) suggests that a holistic view of tourism is necessary to ensure the long-term success of China as an international tourism destination. While it is normal for economic objectives to drive development, sustainable development strategies recognise that, in the long run, economic objectives can only be met if social/cultural and natural resources are maintained. It appears that China's focus was on the economic aspects of tourism during that particular period of time, indicated by the survey results outlined in this chapter. This could lead to the mistaken conclusion that no serious environmental issues were associated with tourism development in China. This, however, was not the case. Given the low level of environmental concern reflected in the residents' perceptions, and the regrettably few studies of the impact of tourism on Chinese local communities, there is a need for more studies from the field to focus on this subject, as recommended in the following.

First, future studies should address the environmental impact of tourism on contemporary China. The fact that the survey respondents concentrated on the economic/social aspects of tourism was no accident, and most likely reflected a serious problem in China's contemporary growth and change – a mass mentality concerned with economic development rather than conservation (Edmonds, 1999; Ma & Liu, 2001; Zhu *et al.*, 2022). Given the size of the Chinese population and the huge potential of China to generate outbound tourists in the future, the environmental impact of tourism growth on China has major global implications. Researchers thus should pay more attention to the environmental impact of tourism on China, in order to help devise appropriate strategies for avoiding the consequences.

Second, research on tourism in China should focus on Chinese traditional ways of viewing nature and landscapes, as well as the implications of this preference for tourism development. Peterson (1995) initiated a theoretical study of this kind, but empirical research is still needed, particularly in terms of how the complex two-way relationship between tourism demand and supply may affect this traditional preference. The results of this type of research will certainly have both local and global significance. Tourism is, in essence, a fashion industry (Prosser, 1994), and the Chinese people's perceptions, expectations, attitudes and values, all of which are essential to their tourism consumption decisions, will be increasingly dynamic as the Chinese economy and society continue to grow and change. Research of this type will not only help local tourism planners adjust their development strategies, but will also provide international tourism operators with insights into appropriate strategies to cope with the emerging Chinese market.

Note

(1) In Chinese, 'yellow' contains a metaphorical connotation of 'promiscuity'.

4 Community Tourism and China's Dilemma of Modernisation

Introduction

The concept of community tourism emerged decades ago, and has been defined and redefined in the context of developed countries. In the last two decades of the 20th century, tourism scholars in China started to apply the concept to the Chinese context. However, there are some essential differences between the perspectives of Chinese and Western scholars regarding the concept of community tourism. The differences are reflected in their different attempts to resolve the following questions: Is community tourism a democratic process or predominantly a planning strategy? Should sustaining community or tourism be the overall objective? To appreciate the differences, this chapter presents the author's study of two specific community tourism developments in China in the first decade of the 21st century. The study took a detour through modernisation and (under)development theories. Attention is directed at the sheer size and diversity of China and its people to consider the social, cultural and economic backgrounds that set China apart from many other less developed nations. In doing so, the study highlights the dilemma of the nation's modernisation campaign, offering a comprehensive understanding of the barriers to exercising community tourism in China.

Having a long feudalist history and under a single-party rule for more than 90 years, China has a highly centralised bureaucracy which, more than in many other less developed societies, exercises a high degree of economic and social control and holds supreme power over national policymaking. With this understanding, the study adopts three major criteria of contemporary modernity (Chamberlain, 1994; Goorha, 2010; Granovetter, 2001; Hollingworth & Boyer, 1997; Jayant & Fahimul, 2004) in order to examine the two cases. The exploration concentrates on the contradictions in China's current modernisation campaign. The study focuses on two cases: Nanshan Cultural Tourism Zone (Nanshan) in Sanya City of Hainan Province and Wan'an Ancient Street (Wan'an) in Huangshan City of Anhui Province (Figure 4.1). They were selected

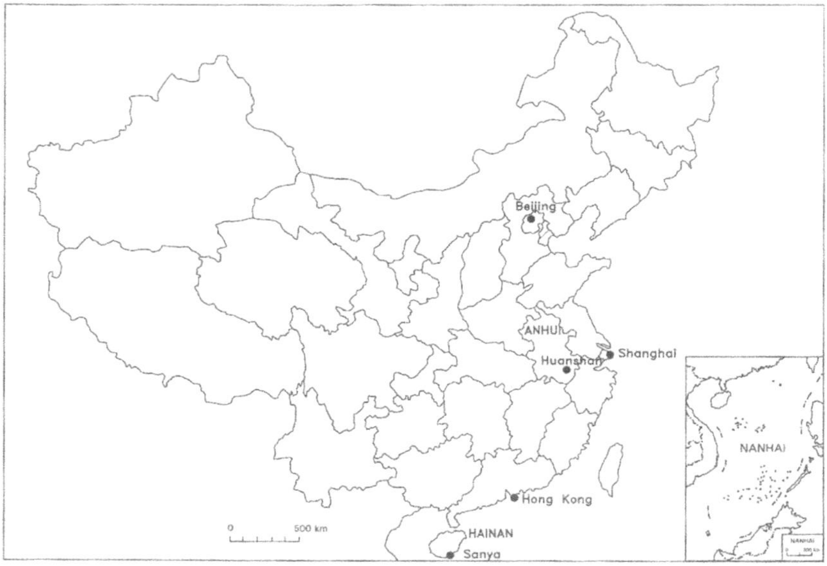

Figure 4.1 Location map of the case studies (Source: Author)

because, firstly, both are popular tourist destinations in China. Secondly, Hainan, once a marginalised provincial backwater, is now the focus of the nation's attention because it is the largest special economic zone (SEZ); while Anhui, historically an important centre of civilisation in China, seems increasingly marginalised by the country's modernisation agenda as it is an inland province. These two destinations, which manifested the tension between China's past tradition and contemporary impetus for 'modernisation', offer some significant insights into understanding contemporary China in general, and community tourism development in particular.

Literature Review

An evolutionary perspective of community tourism

Decades ago, world scholars began the study of community tourism by exploring the impacts of tourism (Butler, 1980; Keller, 1987; Keogh, 1990; Murphy, 1985; Smith, 1989). They suggested that community participation in the tourism development process was crucial to maximising the various socioeconomic benefits while minimising the negative impacts of tourism (Dolezal & Novelli, 2022; Fallon & Kriwoken, 2003; Inskeep, 1991; Ryan & Montgomery, 1994). A sustainable approach to development, namely so-called community tourism or community involvement in the tourism development process, would promote a dynamic tourism–community relationship. Such a relationship could only be established in a modernised civil society where individuals tackle the opportunities and responsibilities of citizenship (Dolezal & Novelli, 2022; Stone, 1989). This concept originated in developed Western countries, and has evolved in both theoretical and practical terms.

The most accepted concept of community tourism is perhaps the well-known advocacy of 'community involvement'. That is, destination communities should get involved in, rather than be excluded from, tourism development within the boundary of the destination area where they are settled. This advocacy can be justified, first of all, by the fact that a destination community creates the basis for community tourism development, because its nature, culture, society and even economy constitute the major, if not all the components, of the tourism products (Hatton, 1999; Horn & Simmons, 2002). Such dependency may vary according to the type of community, but destination communities should not be excluded from participating in the tourism development process. Secondly, it is the destination community that suffers most from the negative impacts of tourism, such as congestion, soaring consumer prices, deteriorating environment and cultural assimilation (Dolezal & Novelli, 2022; George *et al.*, 2009). Therefore, the community must have a say in any kind of development in its own backyard in order to protect its own rights. Finally, if the host community is excluded, conflicts may occur which can have detrimental impacts on sustainable tourism development at the destination level. Arguably, the exclusion of the host community may create increasing antagonistic 'hosts', leading to a 'lose–lose' result for both the community and the industry (Body & Singh, 2003).

The above justification for the advocacy of community involvement seems indisputable. In practice, however, local communities, especially those in rural and peripheral regions, are relatively disadvantaged in terms of education, social class, power, resources and information. Consequently, the benefits of tourism are pocketed by outside operators and the government, while local people must deal with diminished livelihoods (Bell, 1987), and suffer the negative social and environmental impacts even though they are benefiting in economic terms (Dolezal & Novelli, 2022; Scheyvens, 2003). Some destination communities do participate in the decision-making process and share the benefits, yet can exert little real influence on or tangible control over tourism development. In order to reverse such a situation, the stakeholder concept is applied (Anuar *et al.*, 2012; Timothy, 1999; Timur & Getz, 2008) to balance all legitimate voices in tourism destinations. Stakeholders can include local residents, governments and private investors at the primary level; and any party potentially affected by the development at the secondary level (Clarkson, 1995; Li *et al.*, 2020). The importance of each stakeholder should not be determined by her/his possession of critical resources such as money, information and personnel.

Any legitimate voice deserves an equal opportunity to participate in and benefit from the tourism development process. To achieve this, the empowerment of local communities is considered crucial. It appears that both the advocacy of community involvement and the advocates of the stakeholder concept support the participation of local communities in the tourism development process. On the one hand, 'equitable involvement' ensures not only that the opportunity to participate is provided to all concerned in the destination, but also that the benefits are distributed to the population as a whole and not just to individual speculators. On the other hand, only when local communities share equally in the economic, social, psychological and political benefits will they actively support the development. This benign circulation could foster an effective mechanism to sustain both the community and the industry (Mason, 2020). Although it is highly risky to make generalisations about the concept of community tourism because one community is seldom like another, 'equitable involvement' proves to be a very basic concept of community tourism. It has served as a key standard germane to judging if tourism practices in community destinations accord with the spirit of community tourism.

Theorising modernisation and (under)development

Predominantly a Western concept, community tourism is also upheld as one of the so-called 'new directions in tourism' for less developed non-Western societies. Scholars have studied community tourism issues in less developed countries mainly from the perspective of community involvement in economic terms (Blackstock, 2005; Choi & Sirakaya, 2006; Hatton, 1999; Mitchell, 2001a; Timothy, 1999). However, those studies did not address the complexity of applying the concept. Some scholars argue that contemporary tourism development approaches of developed societies may not be directly transferable to less developed countries without considerable adaptations (Tosun & Jenkins, 1998). Their argument is supported by studies on cases from Vietnam and Turkey (Hong *et al.*, 2021; Göymen, 2000; Tosun *et al.*, 2003). They point out that less developed societies would hope to achieve 'modernisation' by prioritising tourism development, thus imagining societies to be in the same capitalist system as developed countries. In reality, however, the less developed world often finds tourism a political as well as a social, economic and moral dilemma, making community tourism an enormously difficult task. Therefore, community tourism and other related issues of the less developed world must be explored in a broader perspective, across the theories of modernisation and (under)development (Harrison, 1988, 1992; Oakes, 1998; Sharpley, 2021; Stronza & Hunt, 2012).

Modernisation theory focuses on the process of industrialisation, whereby the development pattern of industrialised Western countries is proposed as an ideal model for constructing the internal structure of the less developed world. Economically, modernisation is an evolutionary process by which societies shift from agriculture to industry and rural to urban, where money and the money market play a central role. A parallel process, however, is the social aspect where the influence of family and other collectives decline, with institutions becoming more differentiated; and a pivotal role is played by 'modernising' elites and other 'change agents' in introducing modern values and institutions to challenge the resistant tradition. In the process, societal changes occur at both the cultural and psychological levels, with the modern consciousness that allows greater autonomy for the individual.

The modernisation theory is challenged by the (under)development theory which considers development and underdevelopment (as opposed to a lack of development) to be linked elements in the same process. It is argued that development in one part of the world system occurs only at the expense of another part (Dickson *et al.*, 2020; Mandel, 1978). That is, 'centres or metropoles' exploit 'peripheries' or 'satellites' through the mechanism of unequal exchange, thus transferring value from relatively underdeveloped to relatively developed regions. Such an argument explains 'underdevelopment' by reference to the structurally subordinate position of less developed societies within the world system, rather than by the oppressive influence of tradition, the lack of educated elite or the absence of values conducive to capitalist development.

Both theories seem mutually exclusive paradigms, but they have much in common. They are both Eurocentric, embodying the notion of transition from one state to another, and accommodate the idea of a world system – disagreeing, though, on how it is to be envisaged. Both virtually ignore the wants and ambitions of those about to be developed. Guided by either, the welcome of tourism development by the less developed world may appear ambivalent, consequently usually becoming a symptom rather than a cause of economic development. Such a symptom, in one way or another, can find expression in the idea about the paradoxes of modernity.

Modernity and the Chinese perspective of community tourism

Modernity is a tense and paradoxical process through which people produce, confront and negotiate a particular kind of socioeconomic change (Giddens, 1990). As such, modernity has two meanings. One is 'false modernity' constructed in the utopian, teleological modernity of 19th- and 20th-century historicism, of the nation-state and of the institutions of rationalism and scientific objectivity. The other is so-called 'authentic modern' by which modernity is constructed in a process-oriented approach, one in which human subjectivity is ambivalently but irrevocably engaged in a struggle over the trajectory of socioeconomic change. Modernity is thus understood as a condition of paradox, contradiction and basic anxiety that desperately demands, and yet simultaneously denies, constructions of meaning and identity.

Both modernisation and (under)development theories would become rather problematic were they applied to examine the complexity of modernity in China – the development and change associated with the nation's current modernisation campaign. Although China is in the midst of 'modernising' its norms and values, it still has no civil society to counterbalance the power of the central state. Since the Chinese central government decided to open up the country to the outside world in the late 1970s, China has enjoyed the benefits of a growing economy that is generally successful and gradually showing global implications. However, in the view of Choi and Zhou (2001), China is far from being regulated by the invisible hand of the market. There was intense social opposition, especially in the late 1980s when the Tiananmen incident took place, but there have been no major social movements led by 'educated' elites that might introduce modern values and institutions to challenge the current institutional system. China's contemporary development and change are rather complex, and in many ways manifest the paradoxes of modernity.

China's tourism development manifests both 'authentic' and 'false' qualities of modernity that unfolds along with the process of modernisation (Lai *et al.*, 2006; Li *et al.*, 2007; Oakes, 1998). The obsession with achieving an industrialised society remains paramount in China's largely incomplete 'transition' to modernity. In this process, tourism offers an especially appropriate illustration of the paradoxical struggles between the objectifications of 'false modernity' and the promise of an 'authentic' modern subjectivity that is potentially liberating. The experience of current Chinese modernity, therefore, should be perceived as an ongoing struggle for meaning within, along with a desire for repair of, the fragmentation, dislocation and alienation inherent in the modernisation process. Through this, the idea of 'false modernity' emerges as a tempting promise of a resolution to the struggle for meaning and desire for repair. Its primary agents are the nation-state and capital, and one of its principal vehicles is tourism, in which places saturated with tradition and authenticity are constructed and consumed. As such, tourism is marked by the 'misplaced search for authenticity' which, one would argue, is the basis for the Chinese perspective on community tourism.

Around the turn of the century, some Chinese tourism scholars (Li & Wang, 1988; Liu, 1992) began to explore strategies for monitoring tourism impacts which had gradually increased along with the country's modernisation process. Zhang's (1998) 'Translation of Agenda 21 for the Travel and Tourism Industry: Towards Environmentally Sustainable Development' (cf. World Travel and Tourism Council, 1995) indicates that Chinese scholars realised tourism growth could not be sustained by overexploiting the resources and sacrificing the environment. They began to study tourism–community relationships in order to encourage sustainable development (Wu, 2000, 2001). Community involvement thus appeared in the Chinese scholars' working agenda. They recommended that planners should enable community residents to appreciate both the positive and negative impacts of tourism (Liu, 2000, 2001). While trying to prove in theory that tourism needs community support, they proposed a 'stable' tourism–community relationship in line with the Chinese government's top concern of maintaining social and political stability. Their proposal is to sustain tourism rather than the community, explicitly in Liu's (1999: 29) view: 'when the community residents realise they can exercise influence and gain economic benefits, they will support tourism development'. Thus, the essential difference between the perspectives of Chinese and Western scholars about community tourism clearly hinges on the ultimate goal of the practice. For Western scholars, community tourism should be considered a democratic process, with the ultimate goal of sustaining the community rather than tourism or some specific aspects of tourism development. For the Chinese scholars, community tourism remains predominantly a planning strategy with the ultimate goal of sustaining the tourism economy.

Research Methods

A comprehensive review of the literature helps establish a conceptual framework. It highlights that the existing principles of community tourism promote a form of voluntary action by which individuals confront the opportunities and responsibilities of citizenship. The ultimate goal of community tourism is to achieve sustainable development by empowering the destination community. Guided by the principles, community participation in the tourism development process should be regarded as a political as well as a social, economic and moral issue associated with the modernisation process. China, with its long history, sheer size and political system, has experienced a distinct modernisation and development process. To appropriately study a distinctly 'modern' issue such as community involvement in the tourism development process, it is imperative that the complexity of Chinese modernity is examined. In other words, the study focus should be on the contradictions that manifest in China's current modernisation campaign in order to investigate China's community tourism issues. This process of the campaign reflects China's struggles with various contradictions for an 'authentic' modern subjectivity that are gradually revealing China's specific and distinct social realities (Li & Mai, 2022; Oakes, 1998).

Based on the theoretical framework, this study adopts a paradoxical perspective on Chinese modernity, raising two focal questions: What are the major obstacles to actualising community tourism in China? What are the contradictions associated with China's modernisation campaign that have created the obstacles? Three testable hypotheses are proposed for seeking possible answers. The first is that individuals work cooperatively with stakeholder groups to deal with both political and economic issues of mutual concern. This is tested by the first criterion of modernity – the relationship between politics and the economy in Chinese society. The second is that the Chinese institution contains mechanisms for establishing and maintaining cooperation between individuals and powerful stakeholders – the state and entrepreneurs. This is tested by the second criterion of modernity – the question of civil society and the public sphere in the current Chinese society. The third hypothesis is that individuals confront opportunities and responsibilities of citizenship by responding to authoritative decisions that impact on community life. This is tested by the third criterion of modernity – China's development towards democracy. The actual study started in 2001, with empirical field surveys including in-depth interviews and on-site observations. The researcher's understanding of the phenomena in question is developed and grounded on personal data, experiences and analyses (Li, 2002, 2003b, 2004).

Findings at Nanshan of Sanya City, Hainan Island

Located on the south coast of Hainan Island, Sanya used to be a small fishing village (see Figure 4.1). Hainan Island is historically and geographically on the periphery of Chinese civilisation, long reputed for rebelling against the central government and for lawlessness. Through the 1980s and 1990s, China's 'SEZ fever' encouraged the island to separate from Guangdong Province. It became China's largest SEZ in 1988, and the tourism industry has experienced rapid growth ever since (see Table 4.1). Sanya was transformed into a modern tourist city.

Before the 1990s, 40 km west of the city centre of Sanya, the Nanshan area was nothing more than barren mountains and rugged coasts sparsely inhabited by peasants of Li ethnic minority. During the 'SEZ fever', Hainan provincial government selected Nanshan for cultural tourism resort development, on the basis of some legendary tales of Buddhism, which describe Nanshan as a blessed land where the benevolent Guanyin (Goddess of Mercy) vowed 12 oaths to save all living creatures. To dwell permanently in the South China Sea near Nanshan was her third oath. Master Jianzhen, a renowned monk in the Tang Dynasty (618–907 AD), tried in vain five times to sail eastwards to Japan to teach Buddhism. On the fifth sail, however, he drifted to Nanshan where he stayed to preach Buddhism for a year, and successfully reached Japan afterwards. The legendary tales characterise the major themes of Nanshan, such as a 'blessing land of longevity' and 'Buddhist culture'. By the time this specific case study was conducted in 2000, Nanshan had partially opened two sites to the public – Nanshan Temple and Buddhist Cultural Park – with an immediate inflow of tourism receipts.

Table 4.1 Hainan tourism development statistics (1995–2002)

	1995	*1996*	*1997*	*1998*	*1999*	*2000*	*2001*	*2002*
Tourist arrival (10,000)	361	485.8	791	856	929.3	1007.6	1124.8	1254.96
Tourist income (billion RMB)*	52.4	57.23	61.67	66.96	72.46	78.56	87.89	95.38
Annual tourist arrival growth (%)	24.7	34.6	62.8	8.2	8.5	8.5	11.6	8.1
Annual tourist income growth (%)	9.9	9.2	7.8	8.6	8.2	8.4	11.8	7.4
Tourist income of GDP (%)	16.28	14.63	15.04	15.26	15.19	15.20	N/A	N/A

Note: *1 RMB = 0.12 USD.
Adapted from Hainan Tourism Bureau Statistics (2003)

The government state manipulated the development of the resort. The Sanya Municipal Tourism Development Corporation, represented by Sanya City government, had a 40% shareholding in the development agency – Sanya Nanshan Industry and Development Limited (SNIDL) – that involved two other domestic and overseas investors. In China, the government state owns all land rights and determines development strategies as long as residents are compensated. For instance, China's land law (cf. Law of the People's Republic of China on Land Management; China National State Council, 1998) specifies that such compensation must include a land compensation fee, resettlement allowance and compensation for attachments to the land and for young crops. However, in the case of Nanshan, matters regarding land expropriation and compensation issues were negotiated only between the developer (SNIDL) and the landowner (Sanya City government). The residents were excluded. Sanya City government – a shareholder in the development agency – was supposed to share the costs to compensate the residents' relocation. However, the negotiation resulted in a deal that the developer would pay the government's share, while the government calculated the total to distribute to the residents. This encouraged corruption among state officials. Around 2002, the director on board and president of SNIDL was removed due to a corruption scandal which, a few anonymous respondents revealed, had dragged in some higher ranking officials. Since Hainan became China's largest SEZ, corruption scandals within different levels of the provincial government have never gone away (Guo, 2021).

State manipulation of tourism development creates power and money and entices intellectuals into corruption. 'Experts' from prestigious universities and institutions were invited to deliver 'academic' speeches on the relationship between tourism and real estate development. As witnessed at a tourism-related real estate forum, the keynote speeches of the invited 'experts' popularised, with one voice, the theory of tourism–discretionary income relationship by predicting excellent prospects for China's tourism development as the country's commodity shortage was history and that Chinese people were ready to enjoy the 'luxury' of tourism. To a large audience, they speculated promising tourism property investment, and suggested the best bid should be the European-style villas built in Nanshan. However, it was uncovered that the commercial estate was one of many serious problems that emerged in the development of Nanshan at the beginning of the 21st century.

An official from Sanya City tourism bureau described the problems as, firstly, unrealistic market anticipation led to the overexpansion of Nanshan from a planned size of 34.7 km^2 to 50 km^2 and, eventually, to 60 km^2 involving a total investment of RMB70 billion (approximately US$820 million at the time of this study). Neither market potential projections nor a cultural theme were developed, such as the commercial estate, laying the financial burden on the developer. Secondly the developer was unaware that China's religious laws (cf. Regulations regarding Management and Operation of Religious Sites and Temples by China National State Council, 1981) specifies that religious sites should be owned, operated and managed by religious groups. The developer assumed that the Buddhist community would be grateful for their investment, thus taking it for granted that the community would let the developer operate the sites. Once Nanshan started to generate revenue, the Buddhist community claimed ownership. Conflicts thus emerged to hinder further development of Nanshan. Thirdly, the developer's misanalysis of the ecological environment encouraged imports of plants alien to Nanshan – a geologically dry area with little sources of irrigation. Consequently, more than 100 workers needed to be hired to take care of the man-made ecological environment, increasing the financial burden. Fourthly, the representation of Buddhist culture concentrated only on tourism, reflecting the cultural theme rather superficially through architecture such as Nanshan Temple and the giant statue of Guanyin. Near those sacred symbols, European-style villas, luxury hotels and even a golf course were built, distorting the cultural theme and failing to provide visitors with an authentic cultural experience.

State manipulation of the development and its favourable policies towards capital investment also encouraged the entrepreneur to feel superior to the local community. Some managerial staff of SNIDL assumed that the Buddhist community would be grateful for the investment in building the Buddhist sites. They despised the local residents as 'money-driven primitives who know nothing but money' and the Buddhist monks who were 'ignorant of business rules' and 'incapable of handling tourism'. They considered the local residents and the Buddhist monks either 'uneducated and rustic' or 'simple-minded and dumb', incapable of participating in planning the development of Nanshan. They assumed any conflicts with the local community could be easily resolved by 'paying them compensation money'. When asked about the concept of community tourism, their answers were 'I don't know about that sort of thing' or 'it is too difficult and complicated to carry out in practice'.

Nanshan is a remote district of Sanya City. During the time of this case study, the economy was poorly developed and the majority of the residents were poverty-ridden, illiterate or semi-illiterate peasants of Li ethnic minority. In the Duck Pond Village where land was expropriated, it was found that the villagers showed apathy towards participating in Nanshan's development. They regarded tourism as a matter between the *lao ban men* (the entrepreneurs) and government officials, having little to do with themselves. The *Cun Zhang* (chief) of the village, in the interests of the community, initiated contact with the developer in order to get the Li ethnic community involved. But some villagers suspected he had an ulterior motive to gain more compensation money. This upset the chief, and he blamed the villagers' apathy for their narrow vision that holds up to the 'peasant livelihood and mentality'.

The villagers of Li ethnic minority tried to benefit from tourism, rather than develop a realistic understanding of its impacts. Responding to interviews, many of the villagers supported the Nanshan project because it needed their land and they would get compensation money. Few of them thought about how their lives would be changed once the land was gone. The study found that the village lost half of its cultivated land to the Nanshan project, by the time this study started in 2000, while the village population had increased from 680 to 730 people. This was a great challenge to the livelihood of the villagers. The land compensation fee paid to each household ranged from RMB100,000 to RMB300,000 (approximately US$12,000 to US$36,000 at the time of this study) – a large sum of money for the peasants whose average annual household income had never reached RMB3,000 (approximately US$360 at the time of this study). When asked how they had spent or would spend the money, their responses indicated that they would rather use the money to build houses, buy motor vehicles or even consume drugs. Peasants from nearby villages whose land was not to be expropriated were jealous of the 'overnight fortune' of the Duck Pond Village. In order to get compensation, those peasants even lodged a complaint against the developer with the Provincial People's Congress and the Provincial People's Political Consultation Conference.

Findings at Wan'an of Huangshan City in Anhui Province

Huangshan is a mountainous city that covers 154 km^2 of the south corner of Anhui Province (see Figure 4.1). Its name originates from Mount Huang, a mountain resort that is world renowned for its United Nations Educational, Scientific and Cultural Organisation (UNESCO) heritage status. It extends across the four agricultural counties of Yixian, Shexian, Xiuning and Taiping, and three urban districts including

Tunxi city centre, Huizhou urban area and Mount Huang scenic spot. Wan'an, an ancient street of 2.5 km, is located in the agricultural county of Xiuning, and had been the administrative base of the then-prosperous Xiuning County from the period of the Three States (258 AD) to the Sui Dynasty (589 AD). This left a legacy of cultural and economic traditions visible today only in the poorly preserved rows of old stores, workshops and residences along the street, elegantly built in the style typical of south Anhui Province during the old days, reminiscent of a bygone era of glory. An official of Xiuning Tourism Bureau revealed that the county leaders had, for many years, made great efforts to attract investment in order to preserve Wan'an for cultural tourism, but no significant result was ever achieved.

Although the recorded number of domestic tourist arrivals to Huangshan was ranked top among the major cities of Anhui (Table 4.2), tourism development relied solely on Mount Huang scenic spot – a world natural and cultural heritage. The administrative committee of Mount Huang Scenery Development Area (ACMHSDA) has a monopoly on running the scenic spot. Under the administration, Huangshan Tourism Group Corporation – represented by Huangshan City government and Huangshan Tourism Development Limited – and various local businesses and private shareholders, respectively, hold 51% and 49% of the shares in the asset (cf. ACMHSDA, 2003). The general secretary of the Huangshan City Communist Party Committee is the chief director. Authorised by the city government, ACMHSDA operates all fields of the tourism sector including managing and maintaining gardens, hotels, cableways and travel agencies; it has the sole right to prepare materials and allocate funds within the city for tourism development. From 1997 to 2003, the number of subordinate units under Huangshan Tourism Group Corporation increased from 14 to 26. Its total assets increased

Table 4.2 Domestic tourist arrivals at Anhui (as of 2003)

Ranking	*City*	*Domestic tourist arrival 10,000 persons*
1	Huangshen	346.28
2	Hefei	342.94
3	Anqing	333.08
4	Suzhou	235.61
5	Wuhu	227.02
6	Chizhou	203.80
7	Xuancheng	171.75
8	Fuyang	157.35
9	Bengbu	142.23
10	Caohu	137.82

Adapted from School of Geography and Planning (2004).

from RMB43.3 million (approximately US$5.1 million at the time of this study) to RMB107 million (approximately US$13 million at the time of this study). In a similar manner, the number of subordinates under Huangshan Tourism Development Limited increased from 12 at its incorporation in 1996 to 23 by 2003 (cf. Introduction).

The subordinates were mostly concentrated in the three urban districts, running hotels, scenic spot cableways, travel services, florists and gardening as well as entertainment and catering businesses. By 2003, tourist arrivals to Huangshan had reached 6.2 million, generating direct income of up to RMB21.2 billion (approximately US$2.5 billion at the time of this study). ACMHSDA predicted that by 2006, tourist arrivals would reach 10 million per year (cf. Introduction). However, state monopoly resulted in unbalanced developments among the counties and districts. A government official from Wan'an revealed that many heritage assets were neglected in the city's tourism development – outstanding evidence must be Wan'an of Xiuning County.

Xiuning County leaders wanted to preserve this culturally significant ancient street for tourism, but could not do so due to lack of funding. As a lower-level state unit, the county had to wait for the city government to allocate a budget in order to start the project. Additionally, Xiuning is predominately an agricultural county in the city's development agenda, and is thus poor and unattractive to investors. Residents in the scenic spot gain immediate benefits from participating in economic activities such as running family guest houses and souvenir stores. But those from other districts and counties are not even allowed to drive taxis into the scenic spot, according to the ACMHSDA regulations. In 2003, the average annual household income of Tangkou – a village inside the scenic spot district – was RMB22,000 (approximately US$2600 at the time of this study), while that of Wan'an remained below RMB2500 (approximately US$294 at the time of this study) (cf. Xiuning Tourism Bureau, 2004).

Of the 11,000 residents registered in Wan'an at the time of this investigation, nearly half had already left to work in urban areas along China's wealthier coastal cities such as Shanghai and Guangzhou because working in agriculture only created poverty, according to some residents. Those who left could have survived locally to benefit from tourism, but the unfair competition with residents inside the scenic spot only allowed them to work as porters, carrying tourism supplies of 100–150 kg daily, climbing up and down the mountains (average height above 1800 m; a single return foot trip took 10 hours), at a wage of RMB50 (approximately US$6 at the time of this study) a day. So, the local leaders were desperate to exploit their heritage resources for tourism development which would help Wan'an gain a fair share of the city's tourism benefits. However, regarding the community residents' rights to participate in tourism for a fair share of the benefits, the majority of the leaders appeared less supportive.

An official from Xiuning county government addressed the county's need for capital investment which, he emphasised, was vital for the survival of the county's economy. He mentioned that the working agenda of the county government would be simple, namely to ask whoever came to Xiuning from abroad or China's wealthier regions, to bring in investment capital. In order to attract outside investment, the county would guarantee good returns by offering labour three times cheaper than coastal provinces and allowing investors decide how to preserve the cultural heritage of the street. In his opinion, the local people were not qualified to be involved in planning local development projects. Contrary to the hunger for development capital, the local leaders' 'generosity' in spending on 'public relations work' was enormous. During the first five-day field survey in Wan'an, both Wan'an and Xiuning authorities invited the researcher to attend 10 'working' lunches and dinners for 'exchanging ideas' about future tourism development. Usually, two banquet tables of delicacies were served at each meal, costing around RMB2000 (approximately US$235 at the time of this study). Although Xiuning was a poor agricultural county where the average annual household income was below RMB2500 (approximately US$290) at that time, this type of 'public relations work' seemed necessary for the local government. An official of Xiuning County People's Congress held that this was part of Chinese culture, namely 'guanxi' (networks of personal connections) should be cultivated at banquet tables (Li *et al.*, 2007). It was found that the local leaders' 'public relations work' was promptly reported by the 'public' media. One of the banquets made headline news, describing it as 'good work of exercising efforts of both local and international wisdom to help local tourism industry take off'.

Wan'an residents were generally enthusiastic about joining in the tourism ventures of Huangshan in order to share the economic benefits. At the time of this investigation, Wan'an could not offer tourism business opportunities, but many of its residents took part in tourism-related economic activities at the popular tourist sites of the city. These activities included selling handicrafts to tourists and even working as mountain porters. However, involvement in economic activities was limited. Wan'an was famous for winning the gold prize for its locally made compass at the Panama World Fair in 1915. Today, it still produces compasses and sunshine guides as tourist souvenirs. During the field survey, the researcher had the chance to meet and interview Mr A – a Wan'an resident and a successful compass producer. Wan'an authorities arranged the first meeting for a lunch at Mr A's workshop. Among the poorly preserved old houses along the ancient street, Mr A's workshop stood out with its newly renovated appearance. After introductions, the researcher started to feel that the workshop was owned by the county leaders rather than Mr A. He soon left to prepare lunch while the local leaders and the researcher remained in the meeting room to 'discuss' the prospects for tourism development. During lunch, Mr A did not sit at the table to host the guests but rather stood by serving dishes. He seemed happy with his role, always smiling in silence. The leaders seemed satisfied with Mr A's hospitality. They praised his talent of running a good business and promised their continual support. Before the meeting ended, one of the leaders picked up several compasses in the workshop as 'samples' to help Mr A 'promote' the products; he also gave one to the researcher as a 'gift'. Was Mr A happy with the meeting? To what extent was Mr A involved in tourism development? What were Mr A's thoughts about involving himself in the tourism development process? Those questions remained unanswered, thus the researcher decided to return for a personal interview with Mr A.

Three months later, Mr A was interviewed. He admitted that he was probably the sole person in Wan'an who could make a good living from tourism. Making compasses and sunshine guides was a special talent of his family, and he would only pass it down to his descendants. By that time, Mr A owned five stores in Huangshan, selling compasses and sunshine guides to tourists. He made a much better living than other Wan'an residents and was, apparently, contented with the business. When asked about his thoughts on community involvement in the tourism development of Wan'an, he was immediately enthusiastic about participating in the economic activities on the one hand. But on the other hand, he thought it should be up to the government and experts to plan for local tourism development. He believed that ordinary Wan'an residents could not do anything about the specific planning and management of tourism projects. He would be happy as long as he was allowed to do business. He was reluctant to disclose his profit margin or the scale of his business, but his positive attitude towards hosting the lunch three months earlier and his willingness to 'do more of the public relation work' indicated that he had a profitable business. In his opinion, the cost of that lunch was nothing because good returns were assured along with the establishment of 'guanxi'.

Conclusion

This study uncovered a series of contradictions in contemporary China around the turn of the century, which could affect the application of community tourism in the country. First, those contradictions manifest the dialectic relationship between 'centre' and 'periphery' inherent in China's institutional system of economy and politics. Hainan Island, as the first and largest special zone for tourism, plays a prominent role in the nation's modernisation process, despite its historical and geographical peripheral status. Today, Huangshan region, historically one of the civilisation centres of China, is economically marginalised because of its inland location. In the first few years of the 21st century, Anhui was ranked 26th among the 31 provinces and municipalities in terms of economic development potential, while Hainan moved up to 20th place (Yang, 2003). The two selected cases exemplified the tension between the 'centre' and the 'periphery' of China, either historical or contemporary, and the impetus for achieving 'modernisation'. They thus facilitated the exploration of the various contradictions of contemporary China. Through the lens of the three criteria of modernity, this exploration of those contradictions helps identify the following obstacles to actualising the Western concept of community tourism in China.

At the political level, the Chinese nation-state has a wide-ranging and varied reach into every economic sector, forming a major obstacle to community participation in the tourism development process. Such lack of an autonomous economic sphere to counterbalance the nation-state power creates a problem such that market forces and law cannot guarantee the rule of economic rationality, explicit in the making of Hainan's Nanshan project. Through the 1980s and 1990s, 'SEZ fever' characterised China's economic reform. When the Central Committee of the Communist Party formally proposed 'experimental' SEZs in 1979 (Bureau for the Compilation and Translation of Works of Marx, Engels, Lenin and Stalin, 1984), China carved out large tracts of coastal areas for economic development. Hainan was designated as the largest SEZ in 1988, because of its coastal location and strategic proximity to Hong Kong. This transformed Hainan into a tourist island but tourism success was not achieved through rational planning. Initially tourism was not a priority in Hainan's development agenda as the central government's goal was to promote a foreign trade–oriented economy. As the largest SEZ, Hainan in the late 1980s and early 1990s was characterised by a roaring economy based on smuggling, the sex trade and, above all, property speculation, which fuelled corruption.

The provincial government of Hainan had ignored tourism development until it fell into a deep economic malaise two decades later, which forced a radical rethink of the island's development potential as China's warmest and least polluted province. 'SEZ fever' drew the whole nation's attention to Hainan. Waves of migrants came from mainland China to join the 'gold rush for modernisation'. On the island, they found green mountains, crystal clear skies, blue oceans and a simple life. After many years on the cultural and economic periphery of the nation, the island was far behind the modern race for progress. This naturally evoked both nostalgia and fascination – the driving elements behind 'the tourist gaze' (MacCannell, 1989; Urry, 1990; Urry & Larsen, 2011) that pushed the tourism industry to take off. The politics of the Chinese nation-state redefined Hainan, placing it at the centre of the new cultural and economic systems of the modernisation campaign.

In contrast, Wan'an initiated its plan of inviting tourism investment to help its economic survival. Such reality was rather contradictory to its legacy of a 1700-year cultural and economic tradition. Contemporary Chinese nation-state politics almost abandoned the heritage street due to its marginalised geographical location. Although the outstanding Mount Huang scenic spot provides the place with substantial tourism appeal, the nation-state monopoly on tourism practices generates little positive effects on preserving and developing Wan'an. The government should take primary responsibility for resolving the issues regarding the fair distribution of the economic benefits and generating moral guidelines. However, it is rather questionable if the government will be a guardian of justice. A civil society of political pluralism is still out of the question; therefore, that each social group enjoys equal access to opportunities remains a dream. Under such an institutional and political system, it is impossible for the ruling party government to act as a representative of all social groups. If government officials share the same human nature as other people, they cannot simply be expected to behave in a more moral way. Based on this general scepticism about human nature, democratic political processes must be allowed to nurture a public sphere that provides channels for people from every social group to use legitimate means to protect themselves. Otherwise, they will retaliate against the whole society through illegal channels.

Consequently, serious social problems have occurred due to China's contradictions between a rapidly modernising economy and a largely old-fashioned institution. The problems have led to another major obstacle at the business operational level, i.e. to actualising community tourism. The relevant evidence is manifested in both cases. For instance, the business operation is involved with the settlement of social relationships in the exchange of resources among members of society. Since the means of production in China is still largely under state control, there are two types of resources in society – tangible and intangible. Tangible resources are land and materials under state central planning; intangible resources are permission for doing business, certificates for importing and exporting and other opportunities including using one's knowledge and education. The kinds of scarce resources people have and can offer to other people would determine their value in business and in society as well. The education and knowledge possessed by the 'experts' (who delivered the 'academic' speeches at the tourism real estate forum) enabled them to offer information, skills, technology and advice – desirable and fungible resources that are still scarce in China. Their 'brainpower' placed them in the strategic position of the 'business' relationship of resources exchange. Dependent on each other, people with political influence, brainpower and business capital have formed a social elite group to exclude ordinary community residents' involvement in the tourism development process.

During the first three decades of the People's Republic of China, the nation experienced a long period of shortage in development capital due to its centrally planned economy. Thus, in the first 30 years of the nation's opening up, China's market-oriented reform of the economy intended, among other goals, to feed the hunger for capital by diversifying capital sources. In order to attract investment, all levels of government adopted various policies favourable to business developers. This encouraged the growth of an elite business group. The elite's superiority over the masses was manifested in the negative attitude towards community tourism, shown by the managerial staff and the local government officials of both Nanshan and Wan'an. Obscured from public scrutiny, the government can never be a guardian of justice, as indicated by the explicit conflict of interests that overshadowed the negotiation process concerning compensation for the relocated residents of Nanshan. Instead, the government may easily be the source of injustice to fuel corruption, indicated by Wan'an officials' philosophy about 'guanxi'.

'Guanxi' is a network of personal connections. It morally and economically binds members of Chinese society into exchange networks that manifest endless exchanges of banquets and gifts and various economic, political and social benefits among individuals (Li *et al.*, 2007). 'Guanxi', together with other 'traditions' in the Chinese 'culture', creates another obstacle to community tourism at the sociocultural level. Nicely formulated as 'public relations work', Wan'an officials perceived their 'guanxi' networks as the very foundation of society. They could not survive without these networks and must cultivate them at all costs. From the bottom up through all levels of governments, officials attempt to cultivate good relations with their superiors in order to construct horizontal networks vitally important for exercising power and remaining in office. By giving gifts and holding banquets, at the public's expense, government officials attempt to create alliances through cultivating 'guanxi'. Such exercises fuel the corruption joined rather than watched by the 'public' media, explicit in the case of Wan'an. On the one hand, the exercise of cultivating 'guanxi' symbolises a legacy of the single-party rule. On the other hand, it reveals that strong social opposition is still incapable of forming a major social movement towards democracy. Consequently, addressing the issues of economic liberty, economic democracy, political liberty and political democracy is absolutely out of the question – a great obstacle to community tourism.

Other realities stand as obstacles to community tourism. In Nanshan, meeting the basic and felt needs was still the ethnic minority peasants' daily fight for survival at the time of this study. Participating in tourism development demanded time and energy, a luxury the peasants could not afford. As a small ethnic minority group, they had been excluded for centuries from handling social, economic and political issues. It was impossible for them to develop the cultural and economic capacity to handle such issues as tourism development. Due to their peasant livelihood, their interest in participating was very low. They would rather try to gain the benefits of tourism than attempt to determine, through democratic processes, what they could do to achieve the benefits from the development of Nanshan. This type of peasant livelihood, according to Zhang (1957), is die-hard baggage from China's long history of feudalism. It characterises a 'natural economy' conceptualised as peasant production for home consumption by uniting the farming and handicraft industry in peasant households. This led to peasants' lack of vision and their attachment to a singular form of production, resistant to the separation of industry from farming into town workshops and hence to capitalist development.

In contrast to Nanshan residents' inability to involve themselves in tourism, Wan'an residents were enthusiastic about participating. Their understanding of participation, however, was limited to joining in the tourism ventures in order to share the economic benefits. Such economic participation, within the 'guanxi' networks, could hardly contribute to the development of democracy. Given the primacy of interpersonal relations in contemporary China, 'guanxi' has become a major constraint that keeps China from moving towards democracy (Li *et al.*, 2007). The pragmatic functions of 'guanxi' networks and the heavy reliance of the Chinese people on these networks in daily life generate significant costs both in economic and sociocultural terms. Since the resources are under the monopoly control of the nation-state, power remains one of the most important resources. It is chased by all social groups and highly fungible. Different privileges can be achieved as long as one can cultivate 'guanxi' at the right government levels. At the cost of democracy, privileges can be circulated as currency that not only fuels corruption but also creates 'opportunities', which has every indication from the findings of both cases.

Three main arguments thus evolve from this study. First, Chinese nation-state politics exercise supreme power from the top down at different government levels. It manipulates the tourism economy and keeps necessary information from being released to the concerned community residents about how tourism will impact on their lives. If the people concerned have no rights to define their own needs and make their own decisions regarding how tourism should impact on their lives, it is impossible to establish a dynamic tourism–community relationship. Second, community tourism principles are to encourage local residents to question the role of the government in protecting their interests. They must be regarded as a challenge, by the single party that rules contemporary China, to the foundation of the country's large and entrenched bureaucracy. Thus, actualising such principles can be politically risky and requires more bureaucratic formalities that demand more money, organisational skills, time and effort. Applying the Western concept of community tourism to the current Chinese social reality is far from being cost-effective in terms of business operation. Third, the lack of cultural and economic capacity, and the pragmatic norms and values of Chinese society, make the community residents' apathetic towards participation – a vital constraint for actualising community tourism in China.

In sum, China is quite different from industrialised and much more developed Western countries. There is still a long way to go in the country's largely incomplete transition to modernity. The rule of law is far from being in control, democracy is still not on the cards, the market is still not the determining factor in economic life and there still seems to be no intermediary space between society and the nation-state power. The consequences manifested, explicitly with both Nanshan and Wan'an, a series of contradictions. Namely, they are market rules versus political irrationality, development of a modern economy versus construction of a modernised civil society and, traditional Chinese norms and values versus appreciation of a modern sense of democracy. Those contradictions create three key settings – political structural, business operational and sociocultural – in which obstacles stand in the way of actualising community tourism. They encourage the quality of 'false' modernity to prevail in contemporary China.

One would argue, therefore, that the situation in China is still not suitable for the direct application of the Western concept of community tourism. It is important to understand that 'false' and 'authentic' qualities co-exist in Chinese modernity, just like modernity elsewhere in human societies. Without such an understanding, we are unable to appreciate why Chinese tourism scholars hold a different perspective about community tourism. We may mistakenly set China against an ideal world where the market, democracy, law and liberty reign supreme. To assess the applicability of community tourism in China, we must not take for granted that the Chinese economy ought to be giving birth to a new social stratum wedded to the principles of independent individualism. According to modernisation theory, this stratum should arise initially from within the economic sphere through a defence of private accumulation. It should then achieve acceptance at the political level, thanks to its reach into the power apparatus. According to Turner (1993), there ought to be a gradual emergence of a private sphere within which individuals are masters of their own fate; then a public sphere in which matters of common interest can be discussed; and finally democratic procedures will be established to formalise the decision-making process itself. So, it is imperative that appropriate criteria are established to compare China with 'modernised' and much more developed Western societies. The comparison should focus on the possible reasons why China is not, or is not yet, completely 'modern'; or even why in some senses it never can be.

5 Red Tourism and China's Communist Identity

Introduction

So-called red tourism has experienced noteworthy development in China since the last decade of the 20th century. With the support of the nation-state, the heritage of the Chinese Communist Party's (CCP) history, including past revolutionary events, monuments, relics, residences of former communist leaders and other relevant heritage objects, are exploited and promoted for Chinese domestic tourists. For the sole ruling party of China, the CCP, the primary purpose of such tourism practices is to reduce the regional development gap and reconstruct the national identity with the social values and beliefs that had been brought about by the country's economic reform and opening-up policy. The ultimate goal of promoting red tourism is to sustain the rule of the CCP. The Chinese central government is actively involved in both the supply and demand of red tourism. On the supply side, the government plays a significant role in the planning and management process, whereas on the demand side, red tourism consumption is officially organised and certain parts of tourism expenditure are officially subsidised according to the central government's guidelines. Red tourism is therefore politically oriented and manifests certain aspects of general heritage tourism as a means of national identity construction.

Heritage, by its very nature, is selective and value-laden. As a contemporary use of heritage resources, heritage tourism has become an arena of different interests in which various agents are engaged for their own benefits (Schroder-Esch *et al.*, 2006; Zhang *et al.*, 2018). Over the past few decades, heritage tourism has expanded to a form of 'communist heritage tourism' in former 'block' countries of Eastern Europe and red tourism in China. Previous studies in this regard have touched upon the controversial issue of how communist heritage should be represented and interpreted (Bach, 2002; Coles, 2003a; Dujisin, 2007; Hall, 1994; Henderson, 2007; Kim *et al.*, 2007; Light, 2000a, 2000b). Those studies suggest that one of the important characteristics of communist heritage tourism is its ideological orientation, because it presents a period of human history marked with tension between two opposing blocs – socialist countries and capitalist economies. While communism has become a past memory of the former 'block' countries, it continues as a ruling ideology in China for the nation-state to develop the so-called 'socialist country of typical Chinese characteristic'. The study presented in this chapter aims to advance this line of research on communist heritage tourism through an in-depth case analysis of red tourism in China.

Literature Review

The background research begins with a review of heritage and its role in national identity construction to highlight its association with communist heritage tourism in Europe and 'red tourism' in China. This is intended to clarify both the internal and external factors of the tourism system, which account for the development of red tourism in China.

Heritage, tourism and national identity

Heritage includes the cultural and natural environments that people inherit from previous generations. It 'may be viewed as taking on the identity of an interest in the past, an interest in cultures, buildings, artifacts and landscapes of both the past and present' (Boyd, 2002: 212). The significance of heritage in the identity of nations has long been acknowledged. Calhoun (1997) and Smith (2001a) consider heritage to be the shared history of a given society – a fundamental attribute of a nation – because it provides an inexhaustible source of materials that can be utilised to disseminate a sense of nationhood. When presented and interpreted in contemporary institutions such as museums and historical and cultural parks, heritage plays a formative role in building and sustaining a nation-state (Anderson, 1983; Brett, 1996; Peckham, 2003; Zhang *et al.*, 2018). Such potential of heritage in disseminating values, beliefs and norms has been an important factor in the drive of worldwide heritage tourism development.

The significance of heritage tourism as a way of consuming, representing and interpreting heritage is recognised by many. Researchers focus on the management of heritage sites (Aas *et al.*, 2005; Poria *et al.*, 2001), the demand for heritage (Dutta *et al.*, 2007), the possibilities of exploiting heritage for tourism (Li & Lo, 2005), its economic and social benefits (Bowitz & Ibenholt, 2009; Tuan & Navrud, 2008) and sustainable development (Chhabra, 2009; Li *et al.*, 2020). Crucial to heritage tourism is the patronage of tourists. Therefore, tourists' profiles and requirements and their authentic heritage experiences have also become frequent objects of tourism studies (Chhabra *et al.*, 2003; Kim & Jamal, 2007; Poria *et al.*, 2009). In the wave of the worldwide development of heritage tourism, the potential contribution of heritage to building and sustaining the identity of a nation has drawn much academic attention.

Scholars regard heritage as a fundamental attribute of a nation and plays a formative or constitutive role in nation-state building (Calhoun, 1997; Hechter, 2000; Li & Hu, 2008; Li *et al.*, 2010; Mitchell, 2001b; Munasinghe, 2005; Pretes, 2003; Shackel, 2005; Stritch, 2006).

As a fundamental attribute, therefore, the competence of heritage in disseminating values, beliefs and norms shared by a nation is an imperative stimulus for heritage conservation and exploitation. A primary impetus to establish a national park system in the United States, for example, is to recapture the moral quality of the nation's earliest settlers, edify visitors and testify to America's greatness (Mason, 2004). Associated with the worldwide development of tourism, the potential contribution of heritage to national identity construction is increasingly recognised (Evans-Pritchard, 1993; Johnson, 1995; McLean, 1998; Mitchell, 2001b; Munasinghe, 2005; Palmer, 1999; Pretes, 2003; Shackel, 2005; Stritch, 2006). On the one hand, heritage tourism attracts a sizeable domestic and international audience for the dissemination of national commonality. On the other hand, it offers a reliable means to transmit an ideologically authoritative narration of history.

Scholars have long argued for a semiotic relationship between heritage in the tourism context and nationalistic discourse (Culler, 1981; MacCannell, 1976; Li *et al.*, 2010, 2019; Palmer, 1999). Historical buildings, museums, events and natural landscape of national significance are prescribed with mythical discourses and eventually converted into 'metonymic signs' of a nation. Such descriptive vocabularies as 'ancient', 'majestic', 'enduring' and 'powerful', as Palmer (1999) exemplifies, are used to transform heritages into signs, symbols and images of nationhood. By so doing, heritage tourism offers an imagination of the past and is engaged in an ideological framing and reframing of national history to the needs of the nation-state (Johnson, 1999). So, politicians often use heritage and tourism to instil national pride in citizens and impress foreigners (Su & Teo, 2009). So-called heritage tourism is thus 'a tourism of nationalism where the landscape awaiting to be "read" is the landscape of national identity' (Palmer, 1999: 317). Heritagisation of cultural resources, however, is at times intentionally based on an invented, hidden and purposely chosen past (Poria & Ashworth, 2009).

Heritage tourism can be explored as a propaganda vehicle for national identity construction (Palmer, 1999; Pretes, 2003; Tanasescu, 2006; Worden, 2001). National identity represents a generic goal of nationalism (Calhoun, 1997; Guo, 2004; Li & Hu, 2008; Li *et al.*, 2020). It is 'the continuous reproduction and reinterpretation of the pattern of values, symbols, memories, myths and traditions that compose the distinctive heritage of nations, and the identification of individuals with the pattern of heritage and with its cultural elements' (Smith, 2001a: 18). National identity plays a role in bringing together a mix of populations into an 'imagined community' (Anderson, 1983). Forming and maintaining national identity are central to modern nation-state building. As a present use of history, heritage sites, especially those that are deemed of national significance, can be interpreted with a nationalistic discourse for tourist consumption. Heritage tourism thus contributes to integrating a diverse population into a nation (Peckham, 2003). There are many academic studies on the relationship between heritage tourism and national identity construction in the Euro-American contexts and recently in the contexts of Asia-Pacific and Africa (Henderson, 2007; Lepp & Harris, 2008; Michaud & Turner, 2006). Of particular interest in those studies is the trend to situate the relationship in developing societies where the circumstances of politics, the economy and culture are undergoing tremendous changes. As part of this trend, so-called communist heritage tourism in Eastern Europe and red tourism in China have drawn increasing attention from tourism scholars worldwide.

Communist heritage tourism and red tourism

For years, the exploitation of communist heritage for tourism development has been practiced in former communist countries such as Romania, former East Germany and Hungary (Baker, 1993; Coles, 2003a; Dawson, 1999; Light, 2000a; Tunbridge, 1994). After the collapse of the East-European communist regimes in the 1990s, those countries gradually recognised that the material legacies of communism could attract tourists from the West who want to gaze upon communist heritages and experience the history of the communist reign (Light, 2000a). Communist heritage tourism thus started to work as an economic tool to promote the process of social transformation and the privatisation of the tourism industries (Coles, 2003b). Later, however, and not as expected, the social transformation did not benefit the ordinary people, as some citizens of those countries began to embrace die-hard communist heritages with a sense of nostalgia and a longing for a utopian society (Light, 2000b, 2001). For instance, cities such as Berlin and Budapest (Figure 5.1) began to capitalise on their own communist heritage by embedding it in the itineraries of city tours (Light, 2000a). In Albania, communist sites are also successfully included in tourism itineraries (Dujisin, 2007). However, it has also been found that the Romanian people regard the presentation of communist heritage in Romania as an unwanted past (Light, 2000b).

Those former 'block' countries are not alone in attempting to exploit their communist heritage for tourism development. The governments of East Asian countries such as China, North Korea, Laos, Vietnam and Cambodia (Henderson, 2007; Kim *et al.*, 2007) have found that developing communist heritage tourism is a useful tool for enhancing the sense of national identity and propagandising the achievements of the communist administrations (Figure 5.2). For instance, red tourism in China is, on the one hand, regarded as a strategic engine that drives the country's

Figure 5.1 Budapest statue park (Source: Author)

Figure 5.2 Monuments in memorial of late paramount leaders of North Korea (Source: Author)

economic growth, whereas on the other hand, it is promoted mostly to Chinese domestic tourists for the patriotic education of the young generation (Li & Hu, 2008; Li *et al.*, 2010). Red tourism exploits the historical heritage of the CCP, such as revolutionary sites and monuments, residences of former communist leaders and other objects laden with the 'red spirit' – the Chinese communist identity. The so-called red spirit combines communist ideology, patriotism and the traditional Chinese virtue – collectivism (Shao & Wang, 2006). Thus, 'red tourism' should be regarded as a type of heritage tourism because the travels represent the history, identity and inheritance of an area and its people (Nicholls *et al.*, 2004). The primary representation of red tourism includes the communist identity as well as the people who fought and are fighting for the construction, maintenance and sustenance of such identity in both the past and the present (Shao & Wang, 2006).

There are two discernible similarities between communist heritage tourism and red tourism. First, the heritage resources are built on physical legacies formed during the historical period of fighting to establish and construct a socialist nation-state. Second, the symbolic meaning of such heritages is acknowledged as central to attracting tourists. As a present representation of this history, objects for the tourist gaze – monuments, museums, sculptures and martyr tombs – offer opportunities to learn or commemorate communism as a type of ideology and socialism as a polity. A comparison of the two also manifests some distinct features that differentiate red tourism in China from its counterparts in Eastern Europe. First, communist heritage tourism in Eastern Europe is externally driven. Its primary driving force is to meet the demands of foreign tourists, especially those from the West, such as France, the United Kingdom and the United States (Light, 2001). On the contrary, the most active agent promoting red tourism development is the nation-state – the Chinese communist government. Second, the cultural significance of communist heritage tourism and red tourism is quite different. Post-communist governments in Eastern Europe tend to ameliorate their images of communism and reaffirm their status as 'Europeans', which creates the dilemma of 'identity vs economy'. That is, to forget the history of the former communist reign and reconstruct a collective identity of Europeanness on the one hand and to utilise communist heritage for economic development on the other (Coles, 2003b; Till, 2003; Tunbridge, 1994, 1998). In contrast, the Chinese government treats red tourism as a political strategy to stimulate a collective nostalgia for the communist past and balance regional development, so as to justify the legitimacy of the communist reign (Li & Hu, 2008; Li *et al.*, 2010; Shao & Wang, 2006). The connotation of red tourism is thus politics by nature.

However, heritage is not history at all. It is 'not an effort to know what happened but a profession of faith in a past tailored to present-day purposes' (Lowenthal, 1996: xi). Heritage is, after all, 'a selective re-creation and reinterpretation of the past based on contemporary values and ideas' (Olsen & Timothy, 2002: 7). So, fundamental to the dissemination of national identity in heritage tourism is heritage interpretation, because what is actually consumed in heritage tourism is not only the physical items of heritage but also their interpretation (McLean, 1998; Nuryanti, 1996). Heritage interpretation is far more than the supply of factual information and a mere description of physical elements. Its core mission is to educate in an informal manner by broadcasting meanings and messages with interpretive strategies and illustrative media (Staiff, 2014). Thus, heritage interpretation is not necessarily a faithful representation of history, since heritage itself is open to selectivity, value attachment and distortion in the tourism context (Allcock, 1995; Goulding & Domic, 2009). The ways of heritage interpretation thus manifest exclusion and inclusion in representing a selected version of history, which is, after all, to serve specific purposes in trivialising or magnifying certain parts of history with contemporary politics (Lowenthal, 1996).

Three strategies are recommended for interpreting nationalistic discourse in the heritage context (Shackel, 2001). The first is forgetting or excluding the past of subordinate groups, as modern nationalistic movements are always accompanied by conflicts and wars between nationalists and colonists, or between competing nationalistic forces. The history of the winners is legitimatised or centralised while that of the losers is excluded or marginalised in heritage interpretation. The second is creating and reinforcing patriotism. 'The goal of the official public memory is to produce obedient, patriotic citizens' (Shackel, 2001: 659). Labelling themselves as patriots, the ruling group authorises its own historic experience as patriotic and thus memorable, which in turn serves to rally public support, unify patriotic citizens and suppress competing political interest. Consequently, national identity and patriotism are integrated in the narration of national history in many cases. The third is exploiting and interpreting heritage to meet the demand for nostalgia, to release social anomie, indoctrinate citizens and emphasise and justify the historical inevitability of current systems of politics, economy and culture. Those strategies are widely used in socialist countries (Watson, 1994). Nowadays, multiple types of heritage are used to promote the collective identity of a nation, such as national parks (Boyd & Butler, 2000; Shaffer, 2001), religious and vernacular buildings (Evans-Pritchard, 1993; Li *et al.*, 2020; Mitchell, 2001b) and heritage related to national heroes in the form of monuments, sculptures, martyr tombs and battlefield (Chronis, 2005; Edensor, 1997; Johnson, 1995; Shackel, 2001; Tanasescu, 2006). Moreover, subject to the physical and cultural nature of heritages, multifarious dimensions of national identity can be communicated through heritage tourism (Worden, 2001), as in the illustration of how the political, economic and natural facets of American national values are represented at three different heritage sites (Pretes, 2003).

Research Methods

The conceptualisation of the relationship between heritage tourism and national identity construction provides the theoretical base upon which the study of China's nation-state role in red tourism is operationalised. This chapter thus analyses the ways in which red tourism incorporates nationalistic discourses to present and interpret 'red' heritage resources for the purpose of maintaining and sustaining the communist identity in an increasingly open and modernised China. The actual analysis is based on comprehensive documentary research of relevant nation-state policies and an empirical case study. The documentary research focuses on three sources of data: (a) speeches of the CCP leaders and high-ranking government officials, (b) official documents and statistics and (c) Chinese domestic media reports of nationally significant red tourism development projects. Four integral components are scrutinised, namely policy demand, policy decision, policy output and policy outcome. By so doing, the inquiry looks into the factors that entail the factual involvement of the nation-state in general, the specific roles of such involvement in particular and the associated effects on actual 'red' heritage interpretation.

Complementary to the documentary research is the case study on how heritage resources are interpreted in accordance with nationalistic discourses through red tourism presentations. The in-depth case analysis is focused on Ge Le Shan Revolutionary Memorial (GLSRM) in Chongqing City of China. In 2004, the central government designated the city of Chongqing as one of the top '100 classic red tourism destinations', and the city's GLSRM was nominated in 2006 as one of the top '10 red tourism attractions'. From 2006 to 2008, several research trips of at least one month duration were made to the case study site. Face-to-face interviews were utilised to collect the primary data. The informants included officials of the local tourism administration, tourism scholars and industry practitioners. On-site oral interpretations were also recorded with the permission of the tour guides at the red tourism attractions. On-site observation techniques were also used for data collection, targeting the physical environment, interior layout, exhibition objects, photos and other textual materials of the red tourism attraction under investigation. This method of using combined factual evidence from multiple sources can be mutually complementary for securing data triangulation (Campbell *et al.*, 2020; Decrop, 1999).

The four-component framework (McArthur & Hall, 1996) is adopted to search for themes, concepts, messages and techniques of heritage interpretation. Theme refers to the core meaning of heritage, which can be separated into a set of supportive 'concepts'. Themes and concepts are abstracts, articulated and communicated in 'messages' in oral interpretation and interpretive text. Messages give definition and clarity to themes and concepts and lead to greater visitor understanding. Techniques are used to make interpretation more evocable, impressive and accessible. During the empirical survey trips, the researcher attempted to follow the actual visitor itinerary in order to observe and analyse what the visitors might encounter and experience (Campbell *et al.*, 2020; Laws, 1998; Ratz, 2006). The target of analysis included not only the items on exhibition, available activity programmes for visitors and environmental factors that shaped tourist on-site experiences, but also auxiliary services such as parking, ticketing and catering. Using this method was not only to show all the elements that made up the services but also to enhance objective investigation of those elements, thereby identifying the potentials for improving the services and facilities (Laws, 1998). The merits of the aforementioned two approaches were combined and, by doing so, the service flow would provide a useful analytical clue for the researcher to identify all elements of the environment, settings, activities and objects available for tourist consumption.

Research Findings

The comprehensive documentary research primarily concentrated on the analysis of policy demands, decisions and output of red tourism development. This was to scrutinise the economic and social factors that accounted for red tourism development in China.

Economic and social factors for red tourism development

Two factors external to the tourism system were found fundamental to the policy demand for China's red tourism development, including the so-called threat to the leadership of the CCP and the widening gap in economic development among regions of China. At a national red tourism conference, a member of the Standing Committee of the Political Bureau of the CCP Central Committee – Mr Li Chang Chun – gave a keynote speech (2004). He was worried that socialist China might be overthrown by foreign hostile forces if the country followed the Western mode of democracy in the modernisation process. A year later, Zhang (2005) argued that domestic opponents of the CCP would use a 'Western mode' of democracy as a tool to promote bourgeois liberalism in order to confuse the minds of China's young generations. Both believed that China should adopt a series of new strategies, including 'red tourism', in order to develop new patrons, especially among the young generations, of the communist ideology, along with the nation's rapidly changing economic and social environments. Their perspectives led to the recognition of red tourism as a discourse tool that could be utilised to reconstruct the national identity against the odds of change that the country faced in the new century. Thus, red tourism, in essence, recalls the history of the communist revolution and re-informs the Chinese people, especially the young generations, of the CCP's leading contributions to national independence, liberation and prosperity.

Since 1978, China has enjoyed unprecedented economic growth; however, the inland provinces remain underdeveloped even though the coastal regions are becoming increasingly affluent. It is important to note that those underdeveloped regions used to be the cradle of the CCP – the so-called red revolutionary old regions. It is imperative that the CCP adopts red tourism as a strategic tool to cope with the widening gap in development among regions of China. So, red tourism is promoted for the primary purpose of maintaining and sustaining the economic and social stability of China (Han, 2001; Zhuang, 2006).

Four factors are intrinsic to the tourism system and are also fundamental to the policy demand for red tourism development. The first is that the tourism infrastructures in the existing red tourism destinations were poorly developed (Han, 2001; Zhang, 2005). The mountainous terrains of the locations were significant constraints on tourist accessibility and other supporting services such as water and electricity supply. The second is that tourism development of the red tourism destinations was still in its infancy due to the underdeveloped local economies. Services such as catering, shopping, entertainment, accommodation and tourist attractions were basic and insufficient, and could not meet tourist demand (Zhang, 2005). Third was the lack of proper planning for red tourism at the national, provincial and town/city levels. Involved in red tourism, the nation-state has been the major explorer and exploiter of its own heritage for tourism production. This exploitation was undertaken with little integrated consideration of heritage conservation, product design, marketing concept and sustainable tourism principles (Han, 2001; Zhang, 2008). Fourth was the shortage of investment capital. Infrastructure development and planning formulation need substantial financial support. But the economic situation of the red tourism destinations could hardly generate such support. So, financially it was essential at the time that the nation-state be involved in red tourism development.

The follow-up analysis of policy decisions and output uncovered that China's nation-state played specific roles in red tourism – explorer, regulator and coordinator. As an explorer, it initiated projects and directly planned and arranged financial investment in red tourism. Specific action programmes were set for red tourism, including the development of a product distribution system, a transport network plan, a heritage resource conservation plan, a tourism image promotion scheme and a business operation mechanism. Among those, the primary action was to plan a product distribution system across China, through which specifically selected parts of the CCP's history and heritage were identified as the core resources for red tourism development. Twelve thematic red tourism areas were designated for the associated local governments to prepare action plans that would accelerate local red tourism development.

Given that most red tourism destinations are located in underdeveloped inland provinces, the lack of capital funding was a stumbling block to development. To overcome this impediment, specific actions were taken by the central government, including direct financial support and favourable development policies for red tourism in those regions. An estimated RMB1.86 billion (US$1 was approximately RMB6.5 at the time) from the central pool of the budget fund was allocated to support 181 red tourism-related projects for building transportation, power and water supply facilities (China National Tourism Administration [CNTA], 2006, 2009). The central government also encouraged local governments to provide supplementary financial support for specific local red tourism infrastructure construction and heritage conservation. A national red tourism planning document stated that public investment in infrastructure construction should be jointly borne by both the nation-state and relevant local governments (Li, 2004). This was repeatedly echoed in the annual meeting of the National Red Tourism Coordination Executive Team (NRTCET). Private investment was also encouraged. As a result, private capital totalling RMB59.8 million was invested in various red tourism projects, accounting for 43.1% of the total 'red tourism' investment during that period (CNTA, 2006, 2009).

A top-down campaign to promote red tourism was also launched by the Chinese central government. This campaign was supported by all levels of the CCP organisation in order to market red tourism across the country. At a national red tourism conference, the deputy minister of the Propaganda Department of the CCP called for all official media nationwide to join this campaign (Hu, 2005). The nationally influential internet media Xinhua and Renmin and the website of China Central Television were mobilised to upload and publish textual and visual materials of red tourism, to form an internet platform for promoting red tourism. A significant amount of national and local media space was soon devoted to red tourism promotion, with various promotional events (Table 5.1).

All levels of governments were involved with the actual business of red tourism, exercising substantial influence on the operation of tourism attractions and transportation. The China Ministry of Preservation of Cultural Relics, the Ministry of Culture and the Ministry of Civil Affairs joined the nationwide efforts to promote the conservation of communist heritage and market red tourism. Their dual roles as both communist heritage guardians and operators of red tourism businesses

Table 5.1 Red tourism promotional events (as of 2008)

Date	*Venues*	*Events*
April, 2005	Guilin, Guangxi autonomous region	Domestic Tourism Trade Fair 2005
October, 2005	Jin Gang mountain, Jianxi province	Red Tourism Culture Festival
October, 2005	Nanchang, Jianxi province	The First National Red Tourism Exposition
October, 2005	Linyi, Shandong province	China's National Forum of Red Tourism (Linyi)
November, 2005	Xi Bao Po, Hebei province	Xi Bao Po Forum of Red Tourism Development
April, 2006	Chengdu, Sichuan province	Domestic Tourism Trade Fair 2006
June, 2006	Hangzhou, Zhejiang province	China's Red Tourism Festival
June, 2006	Shanghai	The Foundation Ceremony of Red Tourism Sodality
October, 2006	Nanchang, Jianxi province	The Second National Red Tourism Exposition
April, 2007	Changshu, Jiangsu province	China's National Forum of Red Tourism (Changshu)

Adapted from National Red Tourism Coordination Executive Team (2008, 2009).

were to ensure that the nation-state policies regarding red tourism were implemented 'politically right'. The nation-state acted as a regulator by setting rules and regulations for red tourism development. In 2005, the CCP Propaganda Department initiated the National Bases for Patriotism Education, with policy and funding supports particularly for the former red revolutionary old regions to improve red heritage conservation, representation and interpretation. In April 2006, the Chinese central government issued price guidelines for regulating the ticket pricing at red tourism sites, by which all associated businesses must comply with a not-for-profit principle. Accordingly, a 50% discount on the entrance charge was offered to adolescents, full-time students, the People's Liberation Army soldiers and senior citizens. Transportation charter services were planned and offered by the nation-state government to serve tourists during peak seasons. Statistics show that 231 charter trains were arranged in 2006, serving 160,000 tourists. This number increased to 409 charter trains serving 224,000 tourists in 2007 (CNTA, 2006, 2009).

In 2004, a national red tourism planning document was formulated through the joint efforts of the China State Council and the CCP Central Committee, to serve as both political and ideological guidance for red tourism planning. This document outlined the general directions for red tourism development in terms of its political significance, ideological guiding principles, practical business goals and action programmes. The planning was kept strictly in line with the ideas of some nation-state officials who had defined red tourism as a political, cultural and economic strategy for coping with the problems emerging in the process of China's economic reform and opening up to the world (Li, 2004).

China's nation-state acted as coordinator working with different sectors associated with the tourism system, to ensure the success of red tourism development. In April 2005, the team of the institutional arrangement for red tourism – the NRTCET – were formed. Members included all senior officials from 11 central government departments and 3 CCP central departments (Table 5.2). All members were supposed to support red tourism by using their economic, social and political advantages relevant to their affiliated institutions. For instance, the member representing the Ministry of Finance was supposed to arrange for a special fund budget to be allocated to red tourism. Members representing the Ministry of National Railway and Civil Aviation were to ensure that special transportation was properly planned and arranged for red tourism. The NRTCET was also represented by the CCP Propaganda Department to ensure that red tourism was developed on the 'right track' under the nation-state's ideological guidance.

China's nation-state media – Xinhua News – stated that this ever-present government involvement produced 'encouraging' outcomes. The NRTCET released two progress reports in 2007 and 2008. They indicated that in 2007, domestic red tourist visits amounted to

Table 5.2 Institutional arrangement for red tourism (as of 2008)

1. National Red Tourism Coordination Executive Committee	
Central Government Departments	NDRC: National Development and Reform Committee
	CNTA: China National Tourism Administration
	MOE: Ministry of Education
	MCA: Ministry of Civil Affairs
	MOF: Ministry of Finance
	MOC1: Ministry of Construction
	MOR: Ministry of Railways
	MOT: Ministry of Transportation
	MOC2: Ministry of Culture
	CAAC: General Administration of Civil Aviation of China
	BPCR: State Bureau for Preservation of Cultural Relics
Central Departments of CCP	Propaganda Department of CCP Central Committee
	Literature Research Centre of CCP Central Committee
	History Research Centre of CCP Central Committee
2. Provincial Red Tourism Coordination Executive Team	
Provincial Government Departments	Provincial Development and Reform Committee
	Provincial Tourism Administration
	Provincial Department of Education
	Provincial Department of Finance
	Provincial Department of Civil Affairs
	Provincial Department of Construction
	Provincial Department of Transportation
	Provincial Department of Culture
	Provincial Bureau for Preservation of Cultural Relics
Provincial Department of CCP	Propaganda Department of CCP Provincial Committee

Adapted from National Red Tourism Coordination Executive Team (2008, 2009).

0.23 billion visitors, a 17% increase from that in 2006. Tourism income totalled RMB91.7 billion (approximately US$13.5 billion at the time), a 22.8% increase from that in 2006. Red tourism provided 0.37 million jobs in direct employment and 1.44 million jobs in indirect employment. During a time span of three years, the number of domestic red tourism visitors reached more than 0.4 billion, recorded as an 18% annual increase. Red tourism income produced revenues of more than RMB150 billion within three years (NRTCET, 2008, 2011). A survey report suggested that tourists' satisfaction with the red tourism experience recorded 90.7% as 'good' or 'very good'. The report thus concluded that red tourism had been well received by Chinese domestic tourists (NRTCET, 2008, 2011).

Case study: 'Red' heritage interpretation

The documentary research suggests two primary purposes for China's nation-state involvement with red tourism development: (1) reaffirming the legitimacy of the CCP's supreme rule and (2) sustaining the communist identity of the nation. Those purposes are articulated as the nationalistic discourse and incorporated with the specific interpretation of communist ('red') heritage in the tourism attractions, as in the case of Ge Le Shan Revolutionary Memorial (GLSRM). The site of the memorial was a concentration camp that was built by the Nationalist government during China's civil war (1946–1949) for detaining communists and their supporters. The memorial occupies an area of more than 4 km^2, with more than 800 units and facilities. The relics of the concentration camp are preserved to commemorate the communist martyrs and veterans who devoted their lives to overthrowing the former nationalist regime.

The case study uncovered that these interpretation techniques are used for interpreting the communist heritage. First, the historical establishment was painstakingly conserved by planting primeval pines and cypress forests around the museum, which created a very solemn atmosphere at the former concentration camp. For example, Zha Zi Dong Prison (ZZD) – a unit in the camp – was destroyed by the nationalists before they fled to Taiwan in 1949. Based on relevant archive documents, it was rebuilt in the shape of an authentic concentration camp. Second, for tourism operation purposes, exhibition rooms were created out of former prison cells and prison guards' offices. Heritage items such as photos, textual materials and historical objects were displayed in cabinets. Other than the necessary lighting for security and fire warnings, limited supplementary light and audio and visual facilities were installed, Third, live scenes were recreated in exhibition rooms, such as original torture beds, tables, chairs, quilts and cuffs, demonstrating a prison environment as vividly as possible. A dark room was presented with a collection of interrogation devices on display. Fourth, theme exhibitions were arranged to publicise historians' relevant findings. Fifth, a daily interactive night show was put on for visitors, during which they would be 'forced' into the former concentration camp by fully armed 'prison guards' to experience the created 'horror'. These interpretation techniques were employed to portray, stereotypically, two groups of people – the revolutionary communist inmates and their patrons versus the prison guards, revolutionary traitors and nationalist reactionaries.

Central to the interpretation was to demonstrate the heroic profile of the former communist inmates through creating a comprehensive picture of the inmates' prison experiences. The memorial devoted most interpretative space to the introduction of the former inmates and their life in prison. For example, the interpretations portrayed the communist inmates as moral models of 'purity', possessing an optimistic spirit and a positive outlook. They wrote revolutionary letters, essays and songs to overcome the cruelty and horror of prison life. Those of some representative inmates were especially emphasised in both oral and written interpretation texts. In addition to presenting genuinely authentic 'physical' items, the memorial also attempted to interpret the inmates' life using humanitarian approaches, to inform visitors that the communist inmates were ordinary human beings and they could also make mistakes. For example, visitors were told that Mr Xu Xiaoxuan, a heroic leader who was an inmate, made a fatal mistake that resulted in the arrest of many communist members. The administration of the memorial regards the so-called 'demythicising' strategy to be useful in communist heritage interpretation. The memorial is to raise visitors' appreciation of the communist inmates as ordinary human beings. This strategy helps to provide visitors with an 'authentic' experience of the memorial as a former prison, and enhance their respect for the communist inmates' devotion and sacrifice in the construction of communist China and the national identity.

The case study found three main themes regarding 'red' heritage interpretation in the museum. The first is 'sacrifice' that highlights the inmates as fathers, mothers, sons and daughters, who withstood the physical torture in the prison and fought for the communist cause regardless of family affection. The second is 'victimisation' that details the massacre in the concentration camp in 1949. Relevant documentary records, including former inmates' personal memories, are displayed and orally interpreted to further emphasise the inmates' sacrifice for the communist revolution. The third is 'victory' that describes how the inmates withstood inhumane interrogation and physical torture. The interpretation suggests, with special honour and respect, how the inmates' strong will and determination eventually achieved victory over the reactionary nationalists. The last theme is 'loyalty', of which typical examples are that the museum interpretation frequently refers to two statements made by two representative communist inmates: 'My life belongs to the party [CCP], and as long as the party exists, I would never die'; 'Please report to the party [CCP] on the behalf of me, I have done everything the party taught me to do until the last minute of my life'. The case study also found those themes mutually supportive in achieving the purpose of glorifying the inmates as communist martyrs, namely the so-called 'stark interpretation' (Wight & Lennon, 2007) – inclusion and exclusion, memorisation and oblivion are twins in heritage interpretation; by glorifying inmates with 'a stark history', the museum inevitably entails a depreciation of prison guards, nationalist officials and communist traitors.

Conclusion

This study reaffirms that heritage, as the present use of the past, is never value-free. Heritage is drawn on to articulate values, norms and beliefs that are selected and endorsed by different agents of various political interests. The study also suggests that the interpretation of the communist heritage through red tourism is laden with a dichotomy of values, norms and beliefs embraced by communists on the one side and by their opponents on the other side. Inclusion and magnification of communist values, norms and beliefs present the core of the current Chinese nationalistic discourse in cultural and political terms. In the People's Republic of China, patriotism and devotion to the communist ideology are articulated as a dominant connotation of heritage interpretation. This type of communist ideological education has been carried out ever since 1949 and the introduction of the officially manipulated education system. What is new about 'red tourism' is probably that the tourism system is employed as a channel for such ideological education through explicit inclusion of selected parts of the communist history while interpreting and delivering it to the target audience. Such a 'new' method of communist ideology education is not only to cater to the practical market demand from tourists, but also primarily to cope with the sense of crisis that is increasingly felt by the nation-state in a rapidly changing China. As this study indicates, two external factors to the tourism system itself have contributed to the policy demand for China to launch red tourism – the so-called threat to the leadership of the CCP and the widening gap among regions of the nation in economic development.

Economic reform and its open-up policy have enabled China to become a new economic giant on the world stage, but have left an ideological vacuum in its backyard. As China opens up and embraces the world economy, the traditional influences of Marxism, Maoism and other communist ideology decline, especially among the young generations. While China's economy continues to grow, more and more people are becoming anxious about who can 'get rich first' (Li *et al.*, 2007). Therefore, the methods of communist ideology education need to be updated in light of China's ever-changing economic and social environments. A curator of the memorial admitted that today's visitors have no interest in lectures on communist ideology; therefore, the museum must label the ideological and spiritual essence of communism as a common virtue of mankind in order to stimulate their interest. In this vein, a compromised value is embedded in and embodied by the 'red' heritage interpretation, namely communist inmates' loyalty to the CCP and the communist revolution are elevated to the confidence of faith and the persistence of national interest.

'Patriotism' is frequently used to interpret communists and their patrons – a term that eulogises the common value of the Chinese nationalistic spirit. One may argue that this term is too general to reflect any real political affiliation. However, the generality may be of merit in a terminological sense, as it opens the possibility of integrating the communist inmates as a whole, and interpreting an integrated value regardless of their political affiliations. The spirit of persistence with patriotism is articulated through communist heritage interpretation, thereby delivering a message that patriotism is above all other ideals and responsibilities in the minds of the communist revolutionaries. The communist inmates persisted with courage and at any costs. Even though they were imprisoned, separated from family, physically and mentally tortured, they would sacrifice their lives for patriotic purposes without hesitation. The interpretation portrayed the communist revolutionaries as respectable models of patriotism. Throughout Chinese literary history substantial tributes are recorded, generation after generation, to patriots and their patriotic spirit. So, patriotism easily resonates with the domestic tourists. This is the very way that the communist heritage site – GLSRM – has worked to serve the purpose of national identity construction, to integrate all Chinese people as one nation.

Politically, the interpretation of the communist heritage in GLSRM is to justify the legitimacy of the communist rule over China, by creating an image of the CCP as 'saviour' of the Chinese nation. This is evident in the interpretation of the CCP's contribution to liberating the Chinese people from the so-called 'old China' under former nationalist rule. The interpretation highlights the inhumanity of the nationalist regime through the description of the prison guards' brutality. By describing the CCP as a leading force in China's historical fight for national independence and the freedom of the people, the interpretation tends to justify that a socialist nation-state under the leadership of the CCP was the inevitable historical result of China's century-long turmoil of foreign invasions, civil wars and revolutions. GLSRM is thus represented as a metonymic sign of the Chinese nation through red tourism practice, in which the effort of 'making the nation' is inherently engaged in representing and interpreting 'red' heritage. Heritage is thus prescribed with nationalistic discourse in this process, and the past is described as a collective patriotic past, which should be commonly shared, valued and cherished by the whole Chinese nation.

6 The Impacts of the 2008 Beijing Olympic Games

Introduction

Mega-events are big events that are mostly hosted by cities and may have significantly complex and far-reaching impacts on the host cities (Mules, 1998). On the one hand, mega-events provide host cities with opportunities for urban renewal, place marketing, local economy growth and community culture revitalisation. On the other hand, mega-events may create threats to the host cities, such as loss of place distinctiveness, push-off effects and financial risks (Mair *et al.*, 2021). As both the cause and the consequence of globalisation, the roles of mega-events in affecting or reshaping destinations – the host cities – are becoming increasingly important around the world (Roche, 2003). Scholars contend that destination as a place is an important terrain for understanding an episode of globalisation such as a mega-event (Agnew, 1987, 2002; Logan & Molotch, 2007). A destination constantly evolves as an entity, whereas globalisation accelerates the pace of such evolution, and its pattern tends to become increasingly complex (Butler, 2009). It is essential that cities understand the mechanism of destination change in a world full of uncertainty. Accordingly, the ultimate purpose is to formulate management strategies that help guide a destination to develop in the right direction, so as to produce opportunities and mitigate threats (Jago *et al.*, 2010). Thus, there is a need to investigate various factors that affect, positively or negatively, the development process of a mega-event host destination.

Mega-events are considered among the factors that change or reshape destinations. However, the specific roles of mega-events in transforming host destinations are yet to be uncovered. In the mega-event impact literature, few studies have directly assessed the impacts on the host destinations, although a plethora of works do exist examining a variety of mega-event impacts (Carlsen & Taylor, 2003; Essex & Chalkley, 2004; Gratton *et al.*, 2005; Smith, 2001b). However, their foci are the 'host as a city' rather than the 'host as a tourist destination'. Scholars have developed some useful frameworks for assessing the tourism impacts of mega-events (Dwyer *et al.*, 2000; Getz, 1997, 2008), but offer little explanation of such impacts, thus failing to probe the specific characteristics of a tourist destination (Gunn, 1988; Hall, 2005; Smith, 1995). This chapter presents a study that explores the specific roles of a mega-event – the 2008 Beijing Olympic Games – in affecting or changing China's capital city. The study aims, firstly, to advance an understanding of the underlying concept of destination. Secondly, it proposes a conceptual model as a preliminary explanation of the phenomenon in question. Thirdly, it validates the model in a given empirical context.

Literature Review

The term 'destination mix' refers to the 'mixture' of five specific ingredients of a destination place, namely attraction, facility, infrastructure, transportation and hospitality (Fletcher *et al.*, 2018; Inskeep, 1991; Leiper, 1990; Mill & Morrison, 2009). Tourist attraction – the first ingredient – is the most important factor that draws tourists to visit a certain place. Tourist attraction can be further defined in terms of scope, ownership, permanency and drawing power. A destination normally includes attractions such as natural resources, climate, culture, history and ethnicity. The second ingredient is tourist facility which serves as a complementary base for the operation of a destination. Common facilities include lodgings, food and beverages, and supportive sectors such as shops, laundries, guides and recreation centres. Tourist infrastructure, the third ingredient, is composed of all underground and surface structures including water supply, communication networks, health and medical services, power sources, sewage handling and security systems. The fourth ingredient – tourist transportation – refers to a system that consists of four elements: modes, paths, terminals and technology. The fifth is hospitality, which is normally considered the software of the tourism system. Hospitality refers to 'general feelings of welcome that visitors receive while at a destination area... the way that visitor services are delivered... the general feeling of warmth' (Mill & Morrison, 2009: 29). A destination can thus be deemed a mixture of those ingredients, although not every destination must have all of them.

Mega-event destination as a tourist place

For decades, enquiry of place has been one of the building blocks of human geography. The concept of place is commonly believed to denote three sets of meanings in spite of its various interpretations, namely locale, location and sense of place (Agnew, 1987; Cresswell, 2014; Johnston *et al.*, 2000; Massey, 1995; Relph, 1976, 1997; Tuan, 1976). Place as locale refers to a setting and scale for people's daily actions and interactions; place as location refers to a specific point on the earth's surface; and sense of place refers to the subjective feelings people have about places. The concept of place thus can be applied to understanding what a tourist destination implies because a destination is a special place that tourists are drawn to visit. Comparatively, the concept of a destination mix emphasises the 'elements' of a destination; whereas the place concept mostly highlights its 'structure'. Thus, the place concept appears to be able to offer a 'holistic framework' to define a tourist destination. Through a cross-examination of perspectives between tourism research and human geography, a tourist destination can be redefined as a touristic place that consists of three main parts: destination as locale, destination as location and destination as sense of place.

Destination as locale denotes a physical entity similar to a destination mix. Other than hospitality, it may include four elements: tourist attraction, tourist facility, tourist infrastructure and tourist transportation. As location, a destination should take up a certain space on the earth's surface. This can be understood as the 'spatiality' of a destination. In fact, the process of producing spatiality or 'making geographies' begins with the physical body, with the distinctively spatial entity involved in a complex relation with its surroundings (Soja, 2000), whereas the term 'spatiality' has been used to emphasise a certain spatial relationship or connection between humans and their environment. This geographic notion indicates that a destination as location can have two types of spatiality. The first is 'internal spatiality', referring to the spatial relationship among the elements of a given destination mix. The second is 'external spatiality', referring to the spatial location and connection of a destination with other destination(s).

As far as sense of place is concerned, a destination can be defined as visitors', potential or actual, total subjective appraisal of it. In terms of the possible constituents of such appraisal, there exists, arguably, a myriad of mental concepts such as identity, attachment, image, experience, cognitive mapping, authenticity, memory and imagination, which have all been widely studied in a variety of disciplines and fields. For instance, tourism scholars have studied 'destination image' in depth since the 1970s (Beerli & Martín, 2004; Echtner & Ritchie, 1991; Mayo, 1975; Tasci *et al.*, 2007). Previous understanding of destination image points to its actual function as a fundamental constituent of the mental representation of visitors (e.g. Echtner & Ritchie, 1991; Gallarza *et al.*, 2002). Unlike the other perspectives, destination as sense of place is multidimensional because the mental world remains a realm to be explored (Lau & Li, 2015, 2019; Li & Lau, 2022; Stern & Krakover, 1993).

Mega-event impacts on a tourist destination

Mega-events are sometimes labelled 'hallmark events', 'special events', 'spotlight/media events' or 'major events'. Scholars tend to use these terms interchangeably. Therefore, it is difficult to establish a precise definition. Getz (1997: 6) argues that 'the definition of mega-events will therefore always remain subjective. It is really more a question of the relative significance of an event, rather than any particular measure of its size or reach'. Therefore, mega-events are understood in this chapter as 'large-scale cultural (including commercial and sporting) events that have a dramatic character, mass popular appeal and international significance' (Roche, 2000: 1). Event impacts are imposed by the event's ability to involve different ingredients, including people, resources, products, culture, politics and geography. Ritchie (1984) proposes a useful framework of event impacts: the economic, tourism/commercial, physical, sociocultural, psychological and political impacts. Event impacts can be either positive or negative (Hall, 1992), and notably they tend to vary at different event stages, as event can be understood as a process over time (Mair *et al.*, 2021; Roche, 2000).

The above conceptualisation of destination as a tourist place and the impacts of mega-events on a destination offers the possibility of imagining a hypothetical model for evaluating such impacts (Figure 6.1). It supposes three perspectives with regard to the impacts of a mega-event on a host destination. (1) The locale perspective: A mega-event would transform the host destination through enhancing tourist attractions, tourist facilities, tourist infrastructure and tourist transportation. (2) Location perspective: A mega-event would transform the host destination through affecting its internal and external spatiality. (3) Sense of place perspective: A mega-event would transform the host destination due to visitors' subjective appraisal of it (the destination's images change before and after the event). The first two transformations are physical, while the third is mental. What is crucial to note is that 'transformation' is a general term, as event impacts can be positive and negative (Mair *et al.*, 2021; Ritchie, 1984). However, it is a little strong in tone, because

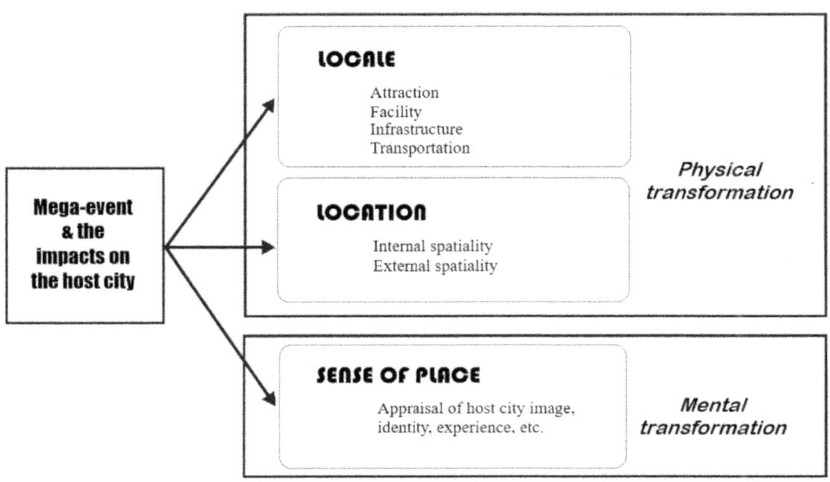

Figure 6.1 Mega-event impacts on a host city (Source: Author)

mega-events nowadays are becoming increasingly large in scale, sufficient to cause evident changes to the host destinations (Essex & Chalkley, 2004; Jago *et al.*, 2010). According to Roche (2000), an important feature of the transforming effects is that such effects tend to fluctuate relative to different stages of a specific event.

Research Methods: A Case Study

The 2008 Beijing Olympic Games provides a suitable exemplary case for testing the validity of the model proposed above. Frigg and Hartmann (2006), Giere (2004) and Silvert (2001) propose the 'Model Theories', considering a model to be a 'presentation' of a certain reality, and can be tested by assessing the model's suitability to that reality.

Evaluating the impacts of the 2008 Beijing Olympic Games

For this specific case study, the model (c.f. Figure 6.1) represents the reality of mega-events' role in transforming the host destinations. In order to test this hypothetical model, it is essential that the extent to which the model matches such a reality is compared and contrasted. The logic of establishing this model grounds on four basic conditions: (1) a mega-event would change the host destination with regard to locale; (2) a mega-event would change the destination with regard to location; (3) a mega-event would change the destination with regard to sense of place; and (4) those transformation effects vary with the stages of the changing process, i.e. before, during and after the event. The model was tested against the empirical context – the Olympic Games that was held in Beijing from 8 to 24 August 2008. This sporting event was chosen because it is a typical mega-event and was regarded as momentous for the rise of China, not to mention its specific features that may offer interesting research findings.

In order to examine the impacts of this mega-event on Beijing, the above four conditions were tested accordingly within the context of the city and the Olympic Games. The first condition was tested by exploring four aspects of the Games' effects on Beijing: attraction, facility, infrastructure and transportation. In doing so, secondary data were used to identify the growth pattern of those aspects from 2001 to 2007. The data sources included the Beijing Statistical Yearbook (BSB, 1992–2008) and relevant planning documents such as Beijing Master Planning (BMG, 2004) maps, Wikipedia and online media reports. On-site participatory observations were also employed to collect information on the event's contributions to such growth. The on-site observations were concentrated during 2007 and 2008, focusing on these places and/or facilities, including the 'Olympic Area' (the Olympic Sports Centre, the Olympic Green and the Olympic Forest Park), Qianmen Street, Wangfujing Avenue, Beijing's subway system, Beijing Capital International Airport,

Beijing–Tianjin High Speed Railway and Beijing–Shanghai Railway. The first condition was met if the Olympic Games were found responsible for the growth of Beijing's destination elements.

The second condition was tested by investigating the Games' effects on both the internal and external spatiality of Beijing as a tourist destination. Specifically, Beijing's subway system was examined. It connects the destination elements of Beijing (at least within the inner city) and if the Games changed the subway system, the Games' effects on the internal spatiality was confirmed. In addition, ground transportation between Beijing and the six Olympic co-host cities – Shenyang, Qinhuangdao, Tianjin, Qingdao, Shanghai and Hong Kong – were examined. The hypothesis that the Games' could generate effects on Beijing's external spatiality was verified if such ground transportations changed. Some Beijing tourist and transportation maps and the co-host cities were utilised as supplementary information for the investigation.

The third condition was tested by investigating the Games' roles in changing visitors' subjective appraisal of Beijing. By doing so, sense of place was operationalised in the destination image because it is virtually impossible to examine all possible mental constructs within a subjective appraisal of a destination, not to mention that the image of a destination can be a representative phenomenon to study when studying the mental world. Mental imaging can be a 'language of thought' or 'mentalese' that underlies any mental state or activity and can serve as a representative of the general notion of sense of place (Aydede, 2019; Lau & Li, 2015; Li & Lau, 2022). An interview survey was employed to collect the relevant data. For the surveys, visitors were invited to freely describe their perceived images of Beijing as a tourist destination. Four rounds of in-depth interviews were conducted between 2007 and 2008, to understand visitors' perceived images of Beijing, and compare the images perceived at different stages during the Games.

Survey No. 1 covered the period between 31 October and 6 November 2007; Survey No. 2 between 17 and 28 June 2008; Survey No. 3 between 8 and 24 August 2008; and Survey No. 4 between 1 and 22 November 2008. A total of 64 visitors were interviewed at Beijing's popular tourist sites, including Tiananmen Square, the Forbidden City, the Summer Palace, Wangfujing Pedestrian Street and Beijing Capital International Airport. Of those interviews, 13 were from Survey No. 1, 17 from Survey No. 2, 15 from Survey No. 3 and 19 from Survey No. 4. All interviews were tape-recorded with the respondents' permissions, and subsequently analysed by CATPAC software (Woelfel, 1998). If the images of Beijing were found different in the investigation period and the Games were responsible for such differences, the third condition was confirmed. The fourth condition was tested by inspecting the starting and ending times of some of the selected projects that were caused, directly or indirectly, by the Games, and by comparing the relative ranks of words related to the Olympic Games in the overall interview texts.

Research Findings

Four key elements of a tourist destination – attraction, facilities, infrastructure and transportation – were developed, consolidated and enhanced in Beijing between 2001 and 2007. The statistics of the selected study period show that Beijing experienced (1) an increased number of key tourist spots; (2) an increased number of star-rated hotels, catering services and other related wholesale and retail venues; (3) the development and enhancement of the infrastructure conditions in the water system, communication network, medical service and health care system, power source, sewage system, streets and security systems; and (4) the consolidation and improvement of the transportation systems including railways, highways and civil aviation (BSB, 2000–2008). The overall growth of those elements was thus confirmed, which helped to identify the Games' contributions (if any) to such growth. Data from on-site observations of some of the relevant sites and from relevant secondary sources also suggest that the Olympics accounted more or less for the identified growth.

Physical transformation: Locale

The study found, firstly, that the Olympic Games enhanced Beijing's man-made attractions. This was confirmed by on-site observations in three places: the Olympic Area, Qianmen Street and Wangfujing Pedestrian Street. Specifically, these enhancements manifested in three aspects: the creation of new tourist attractions, the restoration of old attractions and the reconfiguration of existing attractions. The Olympic Area witnessed a drastic change in land use and the consequent change in its physical appearance. Such a change could be illustrated by inspecting the Olympic Green in the centre of this area. As indicated by Li's (2008) maps, this site had remained by and large on the rural outskirts up to 1996. After it was selected as the site for the main Olympic stadiums, it changed gradually from 2004 onwards, eventually taking on a different shape. This Olympic Area turned out to be one of the most popular tourist attractions in Beijing after the Games. At tourist peak time – for instance, the 'Golden Week' National Day holiday week – the number of visitors to this area even outnumbered those to the traditional tourist hot spots such as the Great Wall and the Summer Palace. Further evidence is the transformation of Qianmen Cultural Street. For decades, it was an important commercial centre in old Beijing, and was substantially renovated by the government of Beijing in preparation for the Olympic Games. Arguably, this renovation project was speeded up, if not motivated, by the Olympic Games, otherwise the street would not have been pushed to reopen on 7 August 2008, on the eve of the Games.

The Olympic Games speeded up the development of tourist facilities in Beijing. In the case of shopping facilities, it was found that the Olympic Games had a bigger role to play. For instance, many Olympic souvenir stores were set up as part of the Olympic commercial programme and those stores proved to be the most popular shops in Beijing. Observation of several such stores in Wangfujing Pedestrian Street indicated that this popularity reached its zenith during the Olympic Games, because every store was flooded with both domestic and international visitors. Some shopping centres were even pushed to renovate because of the Olympic Games, or more precisely because of the opportunity that the Games provided. Take the example of APM, a seven-floor shopping mall invested by Sun Hung Kai Properties Hong Kong. The renovation of the mall in 2007 was mainly triggered by the Olympic Games, as indicated by the 'Introduction' to Beijing APM (Sun Dong An Plaza):

> In the second half of 2005, Hong Kong Sun Hung Kai Real Estate Development Co., Ltd. (hereinafter 'Xindi') purchased 50% of the remaining equity of Xin Dong'an Market from Dong'an group and completed the wholly-owned ownership of the mall. Then, in order to cooperate with the 2008 Beijing Olympic Games, Xindi decided to spend RMB300 million to renovate the shopping malls and office buildings in new Dong'an. At the same time, it renamed the new Dong'an Market Beijing APM.... (APM, 2009)

Observation of the security system in Beijing's subway indicated that the Olympic Games had a moderate role to play in improving the city's tourist infrastructure. In June 2008, a security programme was initiated to ensure the safety of the subway system and the success of the Olympic Games. The programme required that every passenger have her/his carrying luggage examined at a checkpoint before entering the subway station. According to the security staff working in Subway Line 1 and Line 2, this security check system would continue after the Olympic Games and might become a new normal in the future. The Olympic Games also served as an important factor in the improvement and enhancement of the transportation system in Beijing, as confirmed by two observations. One, Beijing's subways system was improved. Although it may be inappropriate to ascribe this to the Olympic Games only, the event indeed accelerated the pace of development. Three lines were added to Beijing's subways in 2002 – Line 8, Line 10 and the Airport Express. Note that Line 8, also named 'Olympic Branch', was built for the Olympic Games (at least by the end of the Paralympics). This is also true for the other two lines, otherwise they would not have been available to operate right before the mega-event began. Therefore, it should not be treated as a mere coincidence. Another case observed was the Terminal 3 project at Beijing Capital International Airport. Documents show that this new project was mainly to serve the Olympic Games, labelled as an aoyungongcheng [Olympic Project]. As Barboza (2008) also observed, Terminal 3 of Beijing Capital Internal Airport was opened on 29 February 2008, as one of the most important supporting facilities of the Beijing Olympics. In the shape of a dragon, the new terminal would serve as a new portal to the world. It was evidently one of the efforts of Beijing's preparation for the mega-event.

Physical transformation: Location

The Olympic Games helped Beijing to become a better organised and more accessible tourist destination, which was manifested in the 2008 edition of the Beijing Transportation and Tourism Map (BMP, 2008). Clearly, Beijing's subway system was greatly improved partially due to the Olympic Games. Perhaps one would argue that the Games at best enhanced the ground transportation. However, taking into account Beijing's transportation and tourism maps, it is clear that this subway network also linked many tourist sites in the inner city (Figure 6.2). By scrutinising this improved subway system, two spatial effects were found.

On the one hand, Line 8, Line 10 and Airport Express – three subway lines built for the Olympics – connect tourist sites such as the Summer Palace, Haidian Park, the Olympic Green (including Bird's Nest and Water Cube), the Olympic Forest Park, Sanlitun Bar and Night Club Street and Chaoyang Park. On the other hand, the updated subway system also enabled Beijing to become a better-connected destination. By connecting to the established subway network (i.e. Line 1, Line 2, Line 13 and Line 5), the new subway network combines the tourist sites alongside both the old and the new subways into a whole system. Figure 6.2 clearly shows that the major tourist attractions, such as Tiananmen Square, the Forbidden City, Wangfujing Street and the Temple of Heaven Park, were interwoven into a more compressed network. In fact, the development of railway transportation changed the methods of travel. Both residents and visitors began to prefer subways to buses/cars for its convenience and punctuality. This is particularly so where Beijing's notorious traffic jams and its geographical massiveness are concerned. Thus, the improvement of the subway system should be regarded as evidence that the Olympic Games enhanced the spatial relationship within the host city itself, although the subway is only one of many possibilities to consider.

The six Olympic co-host cities – Shenyang, Qinhuangdao, Tianjin, Qingdao, Shanghai and Hong Kong – offer proper scope for exploring the Games' impacts on Beijing's external linkage. Four aspects of analysis were operationalised with regard to the geographical locations

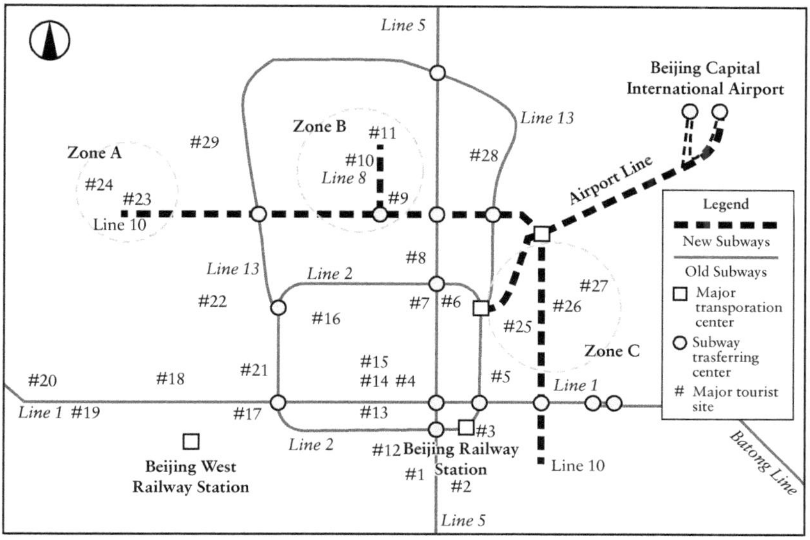

Note 1: Index of the tourist sites in Figure 6.2

#1 Temple of Heaven Park	#16 Houhai Bar Street
#2 Beijing Amusement Park	#17 Capital Museum
#3 The Ming Dynasty Citywall Relics Park	#18 Yutanyuan Park, China Millennium Mouument
#4 Wangfujing Street	#19 Beijing International Sculpture Park
#5 Ritan Park	#20 Shijingshan Amusement Park
#6 Lama Temple	#21 Yuetan Park
#7 Confucian Temple	#22 Beijing Zoo
#8 Ditan Park	#23 Haidian Park (*)
#9 The Olympic Sports Center (*)	#24 The Summer Palace (*)
#10 The Olympic Green (*)	#25 Sanlitun Bar Street (*)
#11 The Olympic Forest Park (*)	#26 Agriculture Exhibition Center (*)
#12 Qianmen Street	#27 Chaoyang Park (*)
#13 Tiananmen Square	#28 Beichen Golf Course
#14 Forbidden City	#29 Ruins of Yuangmingyuan Garden
#15 Jingshan Park, Beihai Park	

Note 2: Tourist sites marked with (*) are those directly affected by the three subway lines that were built for the Olympics. They represents three Zones – A, B, C

Figure 6.2 Beijing's subway system and tourist sites (as of 2008). (Adapted from BMP, 2008)

of those co-host cities. First, the Beijing–Hong Kong tie was not clearly affected by the event. No evidence could be found to show that Beijing–Hong Kong Railway and Beijing–Shanghai Railway were constructed for the Olympic Games. Second, the Beijing–Shanghai tie was not affected either. Beijing–Shanghai Railway began to operate in 1968 and Beijing–Shanghai Express Railway started in April 2008 and would take five years to finish. So, these two connections could not be the result of preparations for the Olympic Games. Third, except with Tianjin, not all Beijing's connections with the co-host cities in the Jing-Jin-Tang Region were improved. The Beijing–Tianjin connection was intensified by the Olympic Games. Beijing-Tianjin Intercity Rail was a project created directly for the Olympic Games. This inter-city rail runs at a speed of 350 km/h, which shortens the 120 km journey between Beijing and Tianjin from 70 minutes to just half an hour. Fourth, the Beijing–Shenyang connection seemed unaffected. One may perhaps argue that Terminal 3 definitely increased the carrying capacity of Beijing Capital International Airport, but no relevant evidence could determine how many visitors would not have come if the terminal had not been built. Despite the difficulty of rejecting the effects of the Olympic Games on Beijing's external spatiality with the six Olympic co-hosts, the effects, if there were any, appeared too weak to be convincing.

Mental transformation: Sense of place

The interviewees perceived the Olympic Games to be responsible for generating new images of Beijing, as manifested in their descriptions of their experiences of the capital city (Table 6.1). Analysing via CATPAC, the salient words in each round of the interview texts suggest that, except for the texts of Survey No. 2, the word 'BeijingOlympics' appeared in the texts of the remaining three surveys. This key word representing the Olympic Games was ranked sixth in Survey No. 1, fifth in Survey No. 3 and fourteenth in Survey No. 4, thus suggesting that visitors agreed that the Beijing Olympic Games had become part of Beijing. If the Games had not been held in Beijing in August 2008, the visitors would not have mentioned this unique word when describing what Beijing meant to them. Thus, this key word could be considered a clear sign in the minds of the visitors that the Olympic Games had influenced the image of Beijing as a tourist destination. Two points should be noted though. First, the totals for the unique words in the four survey transcripts were 458, 332, 294 and 380, respectively. The ranking of these words was determined by the frequencies with which visitors mentioned them to describe their images of Beijing. For instance, in Survey No. 1, the frequency of 'Beijing' was 17.1%, indicating that this word had been mentioned about 78 times (17.1%*458). Words with a frequency above 5% and below 2% were marked in bold. Second, some words were connected, e.g. 'BeijingOlympics', 'ForbiddenCity' and 'TiananmenSquare', to facilitate CATPAC analysis; otherwise, CATPAC would treat them as separate words.

The interviewees not only witnessed the Olympic Games as part of the city and the consequent changes, but they also believed such changes to be positive. Some of the interviewees thought what the Olympic Games offered something 'new' and 'modern' in contrast to the somewhat 'old' and 'traditional' impressions of Beijing. As a Swedish tourist described:

Because we see, for these days, everything about [the Beijing] Olympics. So, we know that Beijing will build for the Olympic Games. But for me, [it is] not the old Beijing. It is the new one. (Direct quote from Survey No. 1; emphasis added)

The above view was shared by an American tourist:

The places [in Beijing] are very beautiful. Like the Great Wall, [it] was obviously incredible to see. And the palaces, the Forbidden City, were really great. They are so much prettier than I can imagine... And it kind seems like a... It's like the old city kind has been really blended with the new city... like you have very old architectural building and something like this [Bird's Nest or Water Cube] which is obviously very modern and

Table 6.1 Top 25 salient words describing visitors' experiences

Survey#1Words	**%**	**Survey#2Words**	**%**	**Survey#3Words**	**%**	**Survey#4Words**	**%**
Beijing	17.1	Beijing	22.6	Beijing	24.8	Beijing	18.7
China	8.2	People	13.6	People	12.6	City	10.5
People	5.6	Places	5.7	City	10.5	People	6.6
Building	4.7	China	4.5	Good	4.8	Really	6.6
City	4.7	Good	4.5	**BeijingOlympics**	4.1	Different	4.5
BeijingOlympics	4.5	Big	4.2	Nice	4.1	Area	4.2
Place	4.2	Chinese	3.9	Big	2.7	Old	4.2
Good	4.0	Different	3.9	Country	2.7	Place	3.9
ForbiddenCity	3.6	Better	3.0	Development	2.7	Time	3.9
TiananmenSquare	3.6	Cities	3.0	Capital	2.4	Good	3.4
Chinese	3.3	City	3.0	Different	2.4	Big	2.9
Old	3.3	Food	3.0	Places	2.4	Culture	2.9
Big	3.1	English	2.7	Politics	2.4	Street	2.4
Built	3.1	Place	2.7	Tourism	2.4	**BeijingOlympics**	2.4
Great	2.9	Visit	2.7	Building	2.0	China	2.4
GreatWall	2.9	Country	2.1	Chinese	2.0	Hutong	2.4
Impression	2.9	Culture	1.8	English	2.0	Olympic	2.4
Beautiful	2.7	Interesting	1.8	Better	1.7	Nice	2.1
History	2.7	Spanish	1.8	China	1.7	Shanghai	2.1
Event	2.2	Time	1.8	Help	1.7	Traffic	2.1
Famous	2.2	Buses	1.5	High	1.7	Visit	2.1
Long	2.2	Countries	1.5	Positive	1.7	Chinese	1.8
Places	2.2	Great	1.5	Visit	1.7	Convenient	1.8
Design	2.0	High	1.5	Change	1.4	HongKong	1.8
Especially	2.0	Hometown	1.5	History	1.4	Architecture	1.6

new. It is interesting to see the really big contrast between that.... (Direct quote from Survey No. 4; emphasis added)

It seems that the effects of the Olympic Games, as a new occurrence, imposed immediate change on the destination images of Beijing. However, it is also necessary to understand that such change was moderate because the core words (e.g. 'Beijing', 'China', 'People', 'Building', 'City' and 'Places' with higher ranks in Table 6.1) adopted by interviewees' descriptions of Beijing were still those that reflected the city itself, its people, the buildings and its position in or implications for China. Nevertheless, the above analysis clearly suggests that, to some extent, the Olympic Games influenced the interviewees' views of Beijing as a tourist destination, adding new elements to the established destination image of China's capital city.

In sum, the case study found that both mental and physical transformation effects varied at different stages of the 2008 Beijing Olympic Games. On the one hand, the effects of the event on Beijing's tourism elements and the spatiality (both internal and external) varied at different stages. Most effects occurred during the pre-event stage, starting in 2003 (two years after the successful bid) and ending in early 2008 (shortly before the Games). This pattern was confirmed by scrutinising some of the changes: (1) the creation of the Olympic Area (including Bird's Nest, Water Cube and the Olympic Forest Park); (2) the restoration of Qianmen Street; (3) the renovation of Wangfujing Street; (4) the building of new subway lines; (5) the construction of Terminal 3 at Beijing Capital International Airport; (6) the renovation of Beijing South Railway Station; and (7) the building of Beijing-Tianjin Intercity Rail (see Table 6.2). On the other hand, the effects of the Olympic Games on Beijing's destination image were also found to vary over time (cf. Table 6.1). Ranking of the word 'BeijingOlympics' was sixth, (n/a), fifth and fourteenth in the four interview surveys, respectively. This indicates that the influence on Beijing's destination images might vary at different event stages, although such difference could not be exclusively attributed to differences in event stage.

Conclusion

Through empirical research, this chapter shows that the 2008 Beijing Olympic Games brought some new content to, or stimulated new developments in the existing destination elements of China's capital city. The mega-event also helped to connect tourist sites in the inner city via an improved and upgraded subway system, and to intensify the link between Beijing and Tianjin – a co-host city. Additionally, the Olympic Games penetrated Beijing's existing destination image, by adding new and positive Olympic elements. Thus, the three conditions for destination transformation – locale, location and sense of place (cf. Figure 6.1) – are supported by the case study of the Beijing Olympic Games. Evidence indicates that the effects of the Games on Beijing varied at different stages

Table 6.2 Time dimension of physical transformation (2003–2008)

Effects Investigated	*Initiated Time (yyyy-mm-dd)*	*Accomplished Time (yyyy-mm-dd)*
The Olympic Area		
Construction of Bird's Nest	2003-12-24	2008-06
Construction of Water Cube	2003-12-24	2008-06
Construction of Olympic Forest Park	2005-12-30	2008-07-01
Qianmen Street Restoration	2007-05	2008-08
Wangfujing Street		
New streetscape	2007	2008
Renovation of APM2		2007
Adding Olympic souvenir stores	2004-08-05	2009-03-31
The Subway System		
Construction of Line 8	2004	2008-07-19
Construction of Line 10	2003-12	2008-07-19
Construction of Airport Express	2005	2008-07-19
Adding security check points	2008-06-29	To continue
Beijing International Capital Airport		
Construction of Terminal 3	2004-03-28	2008-02-29
Beijing South Railway Station		
Renovation	2006-05-09	2008-08-01
Beijing-Tianjin Intercity Rail Construction	2005-07-04	2008-08-01

of the event: the physical transformation appeared to be more evident before the Olympic Games, and the mental transformation tended to witness no clear changing patterns, although slightly stronger during the Games. This supports the fourth condition – the transformation effects varied with the stages of the changing process. The research results suggest that the hypothetical model (cf. Figure 6.1) should be applicable in explaining how a mega-event can contribute to transforming its host city as a tourist destination, in line with the Model Theories (Frigg & Hartmann, 2006; Giere, 2004; Silvert, 2001). Although the term 'transformation' is neutral, it was found to be positive by the specific study of the 2008 Beijing Olympic Games. In this vein, the case analysis of the 2008 Beijing Olympic Games can provide insights into future studies on mega-events and the impacts on host cities.

First, knowledge of mega-events' roles in the evolution of a destination is not static, but unfolds with progress in related research. Previous works on the life cycle of a tourism destination (e.g. Butler, 1998, 2009) indirectly suggest that events are responsible for the destination's change, but offer little explanation about how that actually takes place. Nonetheless, the model introduced in this chapter can be considered such an explanation.

The model divides a mega-event's changing effects on the host destinations into physical and mental transformations. The model validation shows, in the physical transformation, that a mega-event can become part of the host destination, triggering changes in it or redefining its relationship with nearby destinations; whereas in the mental transformation, a mega-event can add new content to visitors' perceptions of a destination. Arguably, a mega-event can be part of the so-called 'dynamism' impacts on a destination's evolution (Butler, 2009). However, Butler's account of events' impacts on a destination's change points predominantly to the negative impacts (e.g. natural disasters, warfare and incidents). This case study of the Beijing Olympic Games' impacts are manifested comparatively more positively, thus it has widened Butler's scope of analysis by linking a mega-event to the change in a tourist destination.

Second, this chapter confirms that the impacts of mega-events are truly complex, flexible and multidimensional per se. This is the fundamental reason why event impact assessments in most cases are incomplete and imprecise (Getz, 2008). As such, a workable assessment framework is necessary for measuring different kinds of event impacts. The model introduced in this chapter can be used as a framework for such purposes. It is proven to be workable in assessing event impacts on destinations, especially in identifying different sub-types of impacts with regard to locale, location and sense of place. It should be noted that this model differs from those offered by other researchers (Dwyer *et al.*, 2000; Ritchie, 1984). Their models are formed by categorising the nature of event impacts into those of economy, politics or environment, whereas the model introduced in this chapter is based on categorising the nature of a tourist destination. Such difference represents the very merit of the newly introduced model, namely serving as a more direct tool for assessing the impacts of mega-events on a destination's change.

Third, the concept of tourist destination needs more comprehensive clarification. This chapter reveals a different understanding of the concept through the perspectives of place and destination mix. This understanding helps reduce the conceptual ambiguity revealed by other researchers (Gunn, 1988; Hall, 2005; Smith, 1995). However, its ultimate conceptualisation (if any) is yet to be facilitated. Within the newly introduced conceptual model ambiguity remains, especially with regard to 'sense of place'. Further exploration of this sub-concept is necessary to advance the relevant understanding. The case study of the 2008 Beijing Olympic Games operationalised the model into destination image, but image is merely a part of the more complex mental world. The extent and the ways of other mental concepts (i.e. place identity, place attachment, experience and memory) should be placed within the conceptual framework of destination (as sense of place), because there remain many tasks to accomplish in this line of research.

7 Leisure Shopping and the Hong Kong–China Relationship

Introduction

Before the so-called 'Individual Visit Scheme' (hereafter Scheme) was launched in 2003, Hong Kong's shopping malls had never seen such huge numbers of patrons. The Scheme was endorsed by the Chinese central government to boost the economies of Hong Kong and Macau, which were devastated by the Severe Acute Respiratory Syndrome (SARS) epidemic that had raged the region in Spring 2003. The Scheme allows individual tourists from China's mainland to visit Hong Kong and Macau. Prior to the Scheme, they could only visit Hong Kong and Macau in group tours. Consequently, shopping malls in Hong Kong have become a popular tourist space that caters to the consumption of tourists from China's mainland.

Statistics on visitors' spending patterns for 2012 indicate that overnight and same-day in-town mainland tourists' total expenditure on shopping was 70.8% and 93.7%, respectively (HKTB, 2013a). Xinhua News (2006), the mainland official media, estimated that 30% of the 65 million shoppers in Harbour City of Hong Kong in 2006 were tourists, of which half were from China's mainland. It has been observed that Hong Kong residents' agitation against China's mainland could be attributed to the overstretching of local resources by mainland visitors (Ip, 2012). Shopping spaces in Hong Kong, as both leisure spaces and tourist resources, have become a contested landscape. The shopping mall, the epitome of tensions across Hong Kong, is thus a key place for understanding the host–guest relationship between Hong Kong local residents and visitors from China's mainland.

This chapter presents research that explores the mutual perceptions of Hong Kong residents and China's mainland tourists when interacting with each other against a contested landscape – shopping malls. The research takes a place theory approach to form the basic argument that the interaction generates a unique type of host–guest encounter – an insider–outsider relationship. Accordingly, tourists may not always be outsiders in the encounter, as delineated in the existing literature of host–guest relationship research. Rather, hosts and guests may have a dual identity, and their mobility may be challenged due to the power struggle within the relationship.

Literature Review

The host–guest relationship is one of the most popular subjects of study in tourism research. However, previous studies predominantly focused on such relationships between Third World hosts and guests from developed countries or between local hosts and foreign tourists. The research in this chapter takes a detour to explore such a relationship in the unique context of 'One Country Two Systems'. From this point of departure, some key concepts and terms are theorised, including shopping, leisure space, landscape and cultural contact, to underpin the overall research.

Shopping mall as leisure space

A contemporary shopping mall is a leisure complex that serves more than consumption. This type of superimposition of a single, massive development creates both opportunities and problems for urban residents (Jackson, 1991).

Shopping is perceived to be a 'social practice of exploration and sightseeing akin to tourism' (Shields, 1992: 102). The purpose of malls could include 'a desire to meet people, feel wanted, exercise, or spend time with friends and relatives' (Timothy, 2005: 22). The meaning of modern malls extends far beyond shopping, because they also function as sites for people to pursue leisure and socialisation (Ahmed *et al.*, 2007; Shim *et al.*, 2013; Stillerman & Salcedo, 2012). Shopping is even regarded as a way of life (Ekici *et al.*, 2018; Erkip, 2003).

Shopping malls are leisure places produced with elements and processes of social and cultural changes (Jackson, 1991), whereas people have a sense of place when they have subjective feelings of a particular space, such as a shopping mall. Thus, people may find individual or group identity in a shopping mall when it becomes part of their daily experiences (Castree, 2003). As contemporary shopping malls evolve, the embodiment of shopping malls has become different. They have become places where people undertake a wide variety of activities, such as window shopping, fine dining, socialising and participating in cultural events (Shim & Santos, 2014; Shim *et al.*, 2013). The shopping mall is thus a place that facilitates contacts for various purposes. Different encounters provide shoppers with various experiences and perceived meanings, shaping the place identity of shopping malls.

Normative landscape and cultural contact

Tourism is multifaceted by nature as both an industry and a sociocultural activity (Li, 2000; Mostafanezhad & Swain, 2019), and a harmonious relationship between hosts and guests is vital to maintaining sustainable tourism development (Griffiths & Sharpley, 2012; Kwong & Li, 2020). This relationship requires friendly attitudes between both hosts and guests, otherwise negative consequences may occur. Therefore, an issue is raised about the cultural differences in the encounters between hosts and guests. During travel, tourists may find themselves in an environment that is culturally different from home due to the cultural differences between hosts and guests. So, the extent of mutual understandings will determine the quality of the interaction, thereby affecting travel satisfaction.

The distinction between 'us' and 'them' may be explained by cultural differences among different nationalities (Hofstede, 2001; Reisinger & Turner, 2003). Since differing national cultures exist in tourists' behaviours, accordingly perceptions of hosts towards guests are informed and shaped (Pizam & Sussmann, 1995). However, it is questionable when an analysis of tourists' behaviours is solely based on nationality (Dann, 1993). Some historical factors may shape different national identities among the same nationals, thus affecting the host–guest relationship in tourism encounters. Some scholars suggest that awareness of nationalism be based on a dichotomy of 'self' versus 'other' (Griffiths & Sharpley, 2012: 2015) and consequently, the host–guest relationship may be prejudiced or impacted by nationalism. The perception of 'us' and 'them' could be further reinforced if the contact is not satisfactory, leading to negative perceptions and even conflicts between hosts and guests (Ap, 2001).

The existing literature offers a myriad of studies on perceptions and attitudes in the tourism context. A geographical concept of 'insideness and outsideness' offers an alternative look into such research on the host–guest relationship. 'The inside-outside division thus represents itself as a simple but basic dualism, one that is fundamental in our experiences of lived-space and one that provides the essence of place' (Relph, 1976: 49). This concept resonates with the notion of 'in place' and 'out of place', namely place is used to construct normative geographies that define appropriate behaviours in a particular context (Cresswell, 1996). A normative landscape is structured in which an ideology about what is right or wrong is transmitted. It is important to realise that sociocultural values are manifested in geographical matters. An individual's behaviour could be 'right' in one place but 'wrong' in another place, according to the place rules. When the rules are violated, transgression occurs, which 'marks the shift from the unspoken unquestioned power of place over taken-for-granted behaviour to an official orthodoxy concerning what is proper as opposed to what is not proper – that which is in place to that which is out of place' (Cresswell, 1996: 10).

Social interaction is inevitably impacted by a variety of social rules and cultural norms during tourism encounters. Therefore, host–guest contact framed by respective cultures may create positive or negative experiences. Tourists' presence and exploitation may trigger hosts' negative attitude towards their guests (Concu & Atzeni, 2012). For instance, hosts consider 'new age travellers' plagues and diseases (Cresswell, 1997); visitors 'cool invaders' (Iyer, 1988); and tourists 'golden hordes' (Turner & Ash, 1975). Given a negative encounter, tourists may also regard themselves as outsiders or intruders (Griffiths & Sharpley, 2012), whereas an authority exists to enforce or strengthen the place rules (Cresswell, 1996), which implies a power struggle within the landscape.

However, the insider–outsider relationship is fluid because the relationship between people and place is dynamic, as is the host–guest relationship. In the tourism context, it is open to negotiation who the 'outsider' or the 'insider' is, and intuitively, such a relationship involves the politics of mobility. Mobility is closely related to accessibility (Stutz, 2006), and it is more than movement from A to B (Jensen, 2014). Mobility is staged partially by social relations, giving meaning to movement. A physical movement bears a meaning or meanings (Cresswell, 2010). The experience of movement is determined by the practice. That is, whether the mobility is a personal choice (Cresswell, 2010). To understand social relations, Cresswell (2010) has developed six elements of the politics of mobility: starting point, speed, rhythm, routing, experience and friction. Experience denotes the feelings of movement; friction concerns the reasons for not moving, the process of stopping and whether the practice is voluntary. Within the insider–outsider relationship, the identities of insider and outsider may create or change the meaning of accessing a place, moving within the place, as well as moving to other places.

Research Methods

The above literature theorisation helps establish the argument that a shopping mall should be regarded as a leisurescape – a normative landscape where place rules and power struggles are present to shape mobility and place identity, transforming the identities of insiders and outsiders. Accordingly, this research explores two specific propositions. First, an individual could bear a dual insider–outsider identity in the same place where a host–guest encounter takes place. Second, the in-place/out-of-place status is not solely decided by place rules, but may also depend on the interest group that wins the power struggle involved with the politics of mobility. A combination of a questionnaire survey and in-depth interviews was employed to investigate the phenomena in question. The population for investigation comprised Hong Kong residents and China's mainland tourists. Hong Kong residents included all residents eligible to obtain a Hong Kong identity card. Mainland tourists were both same-day travellers and overnight visitors, regardless of their mode of visit. Shopping malls as popular attractions where the target respondents could be easily found, thus became the basic fieldwork sites. Self-administered questionnaires were designed by adapting Loi and Pearce's (2012) study on perceptions of generally annoying tourist behaviour, which eventually evolved into a 17-item measurement instrument for analysis. The actual survey was carried out on weekdays and at weekends in October 2013, during the China National Day Golden Week. The questionnaire was distributed at five sites: New Town Plaza, Langham Place, IFC Mall, Times Square and Citygate Outlets.

Purposive sampling was utilised, involving the designation of a group of people as the unit of analysis based on their traits or known attributes (Nardi, 2014). This technique ensures that the sample was 'hand-picked' for the topic grounded on their relevance to the issue studied and existing or privileged knowledge about the topic. For this specific investigation, shopping mall users among Hong Kong residents and mainland tourists were selected as the unit of analysis. Traditional Chinese, Simplified Chinese and English versions of the questionnaire were prepared for data collection. A corresponding version was distributed to the respondents upon request or the judgement of the interviewers. Finally, 561 sets of valid questionnaires were collected, and a total of 10 in-depth interviews were conducted from October to December 2013. The respondents showed diverse sociodemographic backgrounds in terms of age, place of origin, length of residence/visit and occupation (Table 7.1).

Table 7.1 Sociodemographic characteristics of respondents

No	Status	Gender	Age group	Education	Occupation	Personal monthly income (HKD)	Length of residence in Hong Kong
R1	Local resident	Male	20-29	Bachelor	Programme Engineer	20,000-29,999	24
R2	Local resident	Female	20-29	Bachelor	Optometrist	10,000-19,999	23
R3	Local resident	Female	50-59	Junior secondary	Housewife, part-time caregiver	Below 10,000	50
R4	Local resident	Female	20-29	Bachelor	Lead researcher	20,000-29,999	5
R5	Local resident	Female	30-39	Associate	Yoga teacher	10,000-19,999	29
No	**Status**	**Gender**	**Age group**	**Education**	**Occupation**	**Personal monthly income (RMB)**	**Place of origin**
V1	Mainland visitor	Female	20-29	Bachelor	Sales Training Manager	¥5000-¥9999	Huibei Province
V2	Mainland visitor	Male	20-29	Bachelor	Manager (Securities)	¥5000-¥9999	Beijing Municipality
V3	Mainland visitor	Male	20-29	Postgraduate	IT Engineer	¥10000-¥14999	Beijing Municipality
V4	Mainland visitor	Male	20-29	Postgraduate	Lecturer	¥5000-¥9999	Gansu Province
V5	Mainland visitor	Female	30-39	Associate	Bank teller	¥5000-¥9999	Guangdong Province

Each interview was semi-structured with some variations based on interviewees' responses. Respondents were asked about their perceptions of each other in terms of their behaviours and attitudes with concrete examples of circumstances. The interviews usually lasted for 30–90 minutes, during which the conversations were audio-recorded and notes were jotted down simultaneously, unless interviewees refused to adopt either of these practices. Survey data were coded, inputted into SPSS 18.0 and analysed using frequencies, *t*-tests and reliability analysis. Chinese scripts of in-depth interviews were transcribed into English. Similar experiences were coded into standardised phrases in order to ensure coherence for content analysis which helped identify and collate themes and issues.

Findings

The perception statements of the questionnaire survey were categorised into two groups for a total of 17 statements. They were defined as Perception of Behaviour for the first 15 and Perception of Crowdedness for the last 2, to facilitate data analysis and discussion. Cronbach's coefficient alpha was used to test the consistency of the groupings. A reliability

Table 7.2 Reliability analysis of the statement groupings

	α	
Groupings of perception statements	**Hong Kong residents**	**Mainland visitors**
Perception of Behaviour	0.942	0.945
They observe local law and order		
They follow the queue		
They give way/seat to the needy		
They care about hygiene		
They care about personal image		
They speak in a good manner		
They have good table manners		
They use the common facilities properly		
They use the common spaces properly		
They are helpful		
They are patient		
They are polite to other shopping mall users		
They are friendly to local residents		
They are friendly to mainland visitors		
They are friendly to foreign visitors		
Perception of Crowdedness	0.859	0.862
They affect my enjoyment of the shopping spaces		
They cause the shopping spaces to be crowded with people		

analysis was performed in both sub-samples (Table 7.2) where Cronbach's alpha was greater than 0.859, well above the minimum acceptable level of 0.5 (Hair *et al.*, 2010a). Thus, internal consistency was assured to justify the categorisations of perception statements.

Using an independent sample *t*-test, the means of perception were compared between the two samples (Table 7.3), suggesting that the perceptions of Hong Kong residents and China's mainland tourists were significantly different. Evidently, Hong Kong residents held negative perceptions towards mainland tourists while mainland tourists had positive perceptions towards Hong Kong residents. In terms of Perception of Behaviour, Hong Kong residents tended to disagree with or remain neutral about the statements. Hong Kong residents' most negative perception towards mainland tourists was the hygiene issue (Statement d), with about 80% in the categories of disagree and strongly disagree (M1 = 1.95). Over 75% of Hong Kong residents disagreed with Statements b and c, both with a mean score of 2.07. Statement n (M1 = 3.06) marked an exception with over half of Hong Kong residents neutral about this statement and 21.3% agreed

Table 7.3 Mean scores of perception statements

Perception statements*	***t*-test**	
	t	**p-value**
Perception of Behaviour	-29.103	.000
They observe local law and order	-25.653	.000
They follow the queue	-31.571	.000
They give way/seat to the needy	-21.77	.000
They care about hygiene	-31.178	.000
They care about personal image	-24.388	.000
They speak in a good manner	-21.915	.000
They have good table manners	-23.48	.000
They use the common facilities properly	-26.57	.000
They use the common spaces properly	-26.599	.000
They are helpful	-18.543	.000
They are patient	-17.954	.000
They are polite to other shopping mall users	-20.767	.000
They are friendly to local residents	-16.005	.000
They are friendly to mainland visitors	-8.762	.000
They are friendly to foreign visitors	-16.697	.000
Perception of Crowdedness	8.163	.000
They affect my enjoyment of the shopping spaces	5.999	.000
They cause the shopping spaces to be crowded with people	9.198	.000

Note: *Scored on 5-point Likert scale with 1-Strongly disagree, 5=Strongly agree
Significant at the 0.05 level

with it. In the tourist sample, perceptions were skewed to agreement. Hong Kong residents were highly perceived to follow a queue (M2 = 4.14). Similarly, some statements received agreement from over 80% of tourist respondents, including 'they observe local law and order' (M2 = 4.01), 'they care about hygiene' (M2 = 4.10), 'they care about personal images' (M2 = 4.04), 'they use the common facilities properly' (M2 = 4.03) and 'they use the common spaces properly' (M2 = 4.01).

Hong Kong residents tended to agree with the two statements in Perception of Crowdedness, while China's mainland tourists disagreed with them. Of Hong Kong resident respondents, 31.4% rated 'Agree' while about 30% had a neutral stance when responding to the statement 'they affect my enjoyment of the shopping spaces'. Comparatively, 27.9% of tourists rated 'neutral' and another 38% chose the 'disagree' option. A total of 64% of Hong Kong resident respondents agreed or strongly agreed with Statement q: 'They cause shopping spaces to become crowded'; the same portion of tourists chose the 'neutral' and 'disagree' options. This suggests that Hong Kong residents largely thought that mainland tourists caused crowdedness in the shopping malls, which affected their use of the spaces.

Consolidating questionnaire results with in-depth interviews

The above findings are supported by respondents' feedback collected through in-depth interviews. Those interviews showed that Hong Kong residents' feedback was generally negative while tourist respondents held positive views of Hong Kong. However, some locals did admit that the residents were bias against mainland tourists, while some tourists also showed some negative perceptions towards Hong Kong and the people. The respondents' feedback was grouped into positive perceptions and negative perceptions between the residents and tourists in the following discussion.

Residents' positive perceptions

R1[1] admitted that Hong Kong people had presumptions about mainland tourists:

> I think there is some bias due to inadequate understandings. I know about mainland tourists mostly from media portraits of them, but do not have much contact with the tourists… and I think we have cultural differences… I think tourists of good education behave very well. (Direct interview quotation of the study)

R2 and R3 pointed out some exceptions:

> Some tourists are well disciplined. They follow the rules… Those educated are polite. It really depends on the types of tourists I encounter. (Direct interview quotation of the study)

> The educated people are okay, their attitudes are good and behave well, some are very polite. (Direct interview quotation of the study)

R4 told a story of her relatives and friends visiting Hong Kong:

> My relatives and friends did realise that they should behave well when they visit Hong Kong. They try to adapt to or learn local cultures. Some have learnt to hold the door for others when going in or out of shopping malls as they observed from local people... Generally, their attitudes are acceptable to me. (Direct interview quotation of the study).

Residents' negative perceptions

R1 pointed out the bad behaviour of the nouveau riche and some long-rooted political issues:

> Some mainland tourists are very noisy. They cut lines, spit in the mall... Their attitudes are bad, especially to the sales staff; these are their cultures, but only part of the tourists; I am quite annoyed by the nouveau riche. They come to spend a lot and push up the retail prices.
>
> They should be aware of the cultural difference, try to adapt to the local culture of Hong Kong. But some really do not listen, particularly the nouveau riche. When we tell them that they should adapt to our culture such as lining up, they argue with us...
>
> Political issues also exist. They are angry at us when we call ourselves Hong Kong people. They think we should consider ourselves Chinese. And we have different views over Taiwan and Japan issues... (Direct interview quotation of the study)

R2 was concerned about the overuse of spaces and facilities by the nouveau riche:

> They would make places dirty. They do not follow the rules, eat and drink in public areas where eating and drinking are prohibited. They also occupy spaces or facilities, like benches, for a long time, so other shopping mall users cannot use... They make the malls over crowded, and they speak so loudly. Those nouveau riche are impolite, very disturbing. (Direct interview quotation of the study)

R3 had similar perceptions:

> They make public facilities such as toilets dirty, they like squatting in places near big windows, and they usually carry big suitcases that always block the passages... They are imperious. (Direct interview quotation of the study)

R4 also pointed out hygiene problems:

I find sometimes they make places dirty out of their low awareness. This leads to hygiene problems around rubbish bins and toilets... They shop a lot so that they carry many bags. And they really talk loudly, although it is acceptable for me. (Research's direct interview quotation)

R5 told of her personal experiences and observations:

They cut lines, do not follow the order, and fall over other people. Once I saw, during a sale in some shops, they were quite umm...'wild' and made clothes messy in the shops;

I have some observations too. I feel that what the media reported about their misbehaviour such as urinating in open space has reduced a lot. But they really shop crazy, like they are unable to buy those stuff anymore in the mainland. When I see them squatting everywhere, I wonder why they don't sit on benches, which seems to have become a landscape of Hong Kong, and we can't really change anything. (Direct interview quotation of the study)

Tourists' positive perceptions

T1[2] showed very positive perceptions towards Hong Kong people:

They are really friendly and nice. They are patient when telling us the directions, and would even bring us there... We have a very good impression of Hong Kong people – people here have good qualities and are well-educated. (Direct interview quotation of the study)

T2 pointed out:

Compared with people from the mainland, people here are much better in the way that they follow the order... I do not know Cantonese so I use English to communicate in Hong Kong, but not everyone knows English in Hong Kong. But they are still friendly. (Direct interview quotation of the study)

T3 explained the good hygienic conditions from his personal experience:

Local people follow the rules, the hygienic conditions are very good in Hong Kong, and I am impressed. Hong Kong is very clean, so my light-coloured clothes will not get dirty; you know this is almost impossible in the mainland. In terms of their attitude towards tourists, all are very nice and polite. (Direct interview quotation of the study)

T4 agreed with the above comments:

They follow the rules, hygienic; Hong Kong people are generally patient and friendly. (Direct interview quotation of the study)

Tourists' negative perceptions

Only T2, T4 and T5 told of negative perceptions. T5 thought Hong Kong people were sometimes rude to mainland tourists. T2 pointed out the language problem of Hong Kong people. He found that he sometimes could not communicate well with local people either in English or in Putonghua (Mandarin Chinese). T4 also found that Hong Kong people were not proficient in Putonghua.

Behavioural manifestation of residents in the encounter with tourists

Following up residents' negative perceptions, they were further asked how they would behave during an encounter with mainland tourists.

R1 chose to avoid such encounters as much as he could, saying he would leave Hong Kong during long holidays:

> I now avoid going to the malls in Tsim Sha Tsui, because mainland tourists are usually crowded there. I also try to escape Hong Kong, especially during weekends and holidays when mainland tourists come in large crowds. I would go to Asian countries to seek for peace and tranquillity, if I have long holidays. (Direct interview quotation of the study)

R2 chose to enjoy alternate activities for leisure:

> Now I will go to the malls where mainland tourists usually do not go, such as Olympian City in Tai Kok Tsui. But I go to shopping malls less frequently now, especially I would avoid going there during holidays; I can spend my leisure time on other activities, such as jogging, playing tennis or going to theatres. (Direct interview quotation of the study)

R3 simply said that she would rather stay home for leisure:

> I really prefer not to go to malls now; I would rather stay home, because too many tourists are crowded there during weekends and holidays, which is not fun as before. (Direct interview quotation of the study)

R4 pointed out that she just wanted to avoid the crowds:

> I just choose to avoid going to the malls in Mongkok and Causeway Bay, period! There are too many people. It is the crowd that annoys me, I really do not care about their attitudes or behaviours. (Direct interview quotation of the study)

R5 had more to say than the other respondents:

> When I can choose, I do not go with my kids to shopping malls, unless I need to use the toilet or enjoy air-conditioning there. Nowadays the

> shopping mall facilities or selection of goods cannot cater to us Hong Kong people.
>
> I liked shopping or simply window shopping in Causeway Bay very much before, but now when I go there, I feel, wow, so many people, so will want to leave immediately. Now I prefer staying home rather than destroying my leisure time over there. I spend much more time at home now, well, I really feel... the lack of space in the malls.
>
> I can't afford travelling overseas frequently, because it is costly. I can only travel abroad once a year with my whole family. But thanks to the nature of my job, I need to work on holidays, weekends or evenings, so when I have time to go shopping, it is not the peak hour with crowds. (Direct interview quotation of the study)

Conclusion

The survey results indicate that it is the hosts – the Hong Kong residents who play active roles in setting the 'place rules' for their guests – mainland tourists. Those rules concerning what is right and what is wrong are fundamentally collective values shared by the residents. Local people's behaviours are perceived to be the right ones inside the shopping malls where the place rules are in force. In contrast, mainland tourists are perceived to have violated those rules with their deviant behaviours such as neither caring about hygiene nor following the queue. Thus, Hong Kong residents are in place while tourists are out of place. To some extent, tourists acknowledge the existence of those rules by revealing their positive perceptions towards Hong Kong residents, reinforcing the rules and the insider status of the residents.

The survey also shows that mainland tourists were regarded as outsiders in shopping malls in Hong Kong because their behaviour did not match those rules. This leads to a power struggle through which some local residents were forced to leave malls when crowded with mainland tourists. This reality, to a certain extent, transforms the morphology of the malls in Hong Kong. To cope with this change, some residents had to alter their mall visiting habits. All the resident interviewees demonstrated some behavioural adjustments. This phenomenon is similar to the study finding that overuse of resources affects tourism development, and consequently residents have to adopt 'coping mechanisms' in order to avoid direct competition with tourists for leisure resources (Gursoy *et al.*, 2002; Gursoy & Rutherford, 2004).

The residents either had to shift their shopping activities to other malls or choose alternate leisure pursuits. This new 'normal' raises a question: who are now out of place? In reality, the politics of mobility exists in sharing leisure spaces between residents and mainland tourists, particularly the facets of experience and friction as suggested by

Cresswell (2010). One would argue that the residents are mobile but immobile outsiders. In the broad level of leisure-seeking, the residents have become more mobile. China's mainland tourists who are conceptually regarded as 'outsiders' have become a factor, with their large crowds and behaviours, to influence mobility.

In the level of using the mall spaces, the residents were relatively less mobile. They believe the shopping spaces have become more crowded with mainland tourists, which is somehow supported by the tourists' responses to the questionnaire survey. The majority of them chose 'neutral' or 'disagree' when responding to the statement that 'they (Hong Kong residents) cause the shopping spaces to be crowded with people'. Additionally, the residents' experience of shopping malls was no longer pleasant, as revealed from the highly rated statement: 'They (mainland tourists) affect my enjoyment of the shopping spaces'. The crowds have made the shopping malls less 'accessible' for the residents, and have become the 'friction' that has stopped the residents from leisure shopping in those malls.

Hence, this power struggle exists in the insider–outsider statuses as well as the host–guest relationship. The new insiders – mainland tourists – are twisting the place rules of a host context, whereas the residents no longer find shopping malls 'appropriate' places for leisure activities. The residents become outsiders of some specific spaces within a place that they call home. They are indeed out of place in the mall context. Therefore, the propositions (set out in Research Methods) are supported: host and guest can bear the dual identity of insider and outsider in one place and the identity status is dependent on the interest group that wins the power struggle involved with the politics of mobility.

Thus, this research suggests that the in-place/out-of-place dichotomy may partially explain the tension between Hong Kong residents and China's mainland tourists. Many of the complaints by netizens about the unacceptable behaviours of mainland tourists in the public spaces of Hong Kong have been posted to Hong Kong local social media (AFP News, 2013; Apple Daily, 2010). Such behaviour of mainland tourists is generally perceived to be proper or acceptable in China, but may offend Hong Kong local laws and norms in places such as public transport, cinemas and parks. It is the capacity of Hong Kong residents to experience such a cultural shock that determines if these anomalies of the tourists could induce conflicts. Such capacity determines how much the hosts appreciate the guests' cultures. Obviously, mainland tourists' cultures are not perceived appropriate compared with the local culture of Hong Kong. Consequently, the tension is further exacerbated by the cultural differences inherited from the one-century British colonial rule, since Hong Kong culture has been bred from a blend of Chinese and Western cultures and values. This very difference has rooted the differentiation between 'us' and 'others' or 'Hong Kong people' and 'mainland Chinese' (Ko, 2012).

In sum, this chapter has shown the dynamic identities in the host–guest relationship despite the unclear division between insideness and outsideness feelings. The chapter shows that the power struggle within the host–guest encounter creates a politics of mobility, increasing the intricacy and ambiguity of the place identity and, in turn, the tourist–host relationship. The case presented in this chapter manifests the relationship among nationals with disparate senses of nationalism. Under the 'One Country, Two Systems' principle, cultural differences between Hong Kong residents and mainland tourists are evident in the host–guest relationship despite the fact that both are Chinese nationals. This informs the tourism literature that nationalism may be one of the significant influences of potent host–guest interactions.

This chapter emphasises the nationalism factor in an intra-national tourism context, which is worthy of a lot more research. Further studies on both intra- and inter-national tourism contexts are needed in order to justify nationalism as an explanatory factor for the dynamic nature of the host–guest relationship (Griffiths & Sharpley, 2012). This chapter has thus contributed to the field of tourism studies through its exploration of the host–guest relationship from the perspective of moral geographies. Tourism is closely linked with people's everyday lives. Tourism has the power to transform the meanings and values of urban places where leisure travel activities are undertaken, as well as the identity and mobility of both the hosts and guests. In tourism, there is great potential to further explore the role of leisure travel and holidays within people's lives and to integrate these discussions into contemporary cultural and social geography (Squire, 1994b).

All in all, the shopping mall represents one of the many urban contested landscapes in which tensions and conflicts between different groups of users exist. This chapter embodies only one battlefield of tensions; other leisure contexts which have also become popular among both tourists and locals are possible sites of investigation. The analysis that uses the concepts of insideness–outsideness and in-place/out-of-place should be extended to a wider scope of relationships including different stakeholders in tourism such as merchants in and outside the tourism industry and government. This type of research will help garner a more nuanced understanding of the interactions in the urban tourism setting.

Notes

(1) 'R' represents 'Resident', thus R1 refers to Resident 1.

(2) 'T' represents 'Tourist', thus T1 refers to Tourist 1.

8 Island Festivals and Sense of Place: The Hong Kong Experience

Introduction

Islands are one of the most popular tourism destinations given their unique attributes of geographical finiteness, sense of distance and environmental insularity (Cave & Brown, 2012; Sharpley, 2003). The World Travel and Tourism Council (WTTC, 2021) reports that the top 10 sovereign states that contribute to gross domestic product through tourism are all island destinations. The development of tourism on islands has elicited interest among scholars across disciplines. However, research on island tourism has predominantly focused on planning and development strategies (Croes & Semrad, 2015), policymaking (Sharpley & Ussi, 2014), economic benefits and sustainability (Ng *et al.*, 2017; Su *et al.*, 2017), marketing (Carlsen & Zulfa, 2016) and host–guest interactions (Cave & Baldacchino, 2012). Those studies mostly adopt the cost–benefit approach, thus requiring holistic investigations of implicit tourism–place relations. An analysis of the cases in this chapter is in response to the call for a new impetus to establish a dialogue between transnationalism and other key concepts such as 'borders', 'translocality' and 'the state' (de Jong & Dannecker, 2018). It offers a timely intervention to understanding the dynamic role of island festivals in transforming one of the world's most popular tourism destinations – Hong Kong.

The number of tourist arrivals to Hong Kong reached 55.91 million in 2019, and tourism-associated expenditure reached HK$256.22 billion (Hong Kong Tourism Board [HKTB], 2019). Most tourist visits are traditionally concentrated in urban areas such as Hong Kong Island and South Kowloon Peninsula. A recent trend in the world tourism industry indicates, however, that the manner in which tourists consume places extends beyond the consumption of local goods and services in urban places (Urry & Larsen, 2011). For instance, the appeal of small outlying islands in Hong Kong has drawn increasing numbers of tourists, particularly to local festivals such as the annual Bun Festival of Cheung Chau Island.

This chapter explores two festivals in Hong Kong to provide understandings of implicit tourism–place relations: one is the Cheung Chau Bun Festival (CCBF) – a community-initiated homegrown festival held annually in Cheung Chau Island; and the other is the Hong Kong Wine and Dine Festival (HKWDF) – a transplanted global celebration held annually in Hong Kong Island. It adopts a comparative approach to contrast their differences in order to identify what makes up the unique features of Hong Kong identity. From a geographical perspective, tourism representations are regarded as skewed snapshots of reality, hence the exploration is focused on the possibility that island community festivals can make or remake people and places. The ultimate goal is to explore the question to what extent and with what characteristics the perceived sociocultural benefits of festivals produce a sense of place (SOP) and its respective locality.

Literature Review

Previous research on islands mostly adopted a cost–benefit approach, neglecting the human–place interaction. This chapter first responds to scholars' call for establishing a dialogue between 'island' and other key concepts such as 'festival', 'place' and 'tourism' to develop a comprehensive understanding of the dynamic role of festivals in transforming island places.

Festival and place

Festival represents community values, ideologies and identities (Lin & Lee, 2020). When people participate in a festival, they are, in fact, experiencing a place and establishing SOP in the community. The conceptions of human-to-human and human-to-resource activities held at different times and spaces explain how people connect with a place and other people when they participate in a festival. Scholars (Allen *et al.*, 2011; Stevenson, 2019) find that the role of festivals becomes an expression of human activity that contributes to the social and cultural life of a place. A festival is a part of the shared identity of a group of people at a particular time and place (Campelo & Aitken, 2011). Cultural festivals are perpetual productions of social relations, offering cultural exchange opportunities (World Tourism Organisation [WTO], 2012), creating identification and establishing SOP.

The humanistic perspective considers a place through a holistic lens based on the analysis of a 'range of human activity and social/psychological processes' (Sampson & Goodrich, 2009: 902). Meanings could be shaped by intangible specifics such as a festival experience in a place. For instance, interacting with local people and savouring local food during festival feasts encourage dialogue between tourists and the local community through cultural exchange, thereby enabling tourists to establish their SOP. SOP is the ability to recognise different identities of a place that are socially and culturally constructed (Relph, 2001). However, place increasingly becomes placelessness that 'manifest wherever human-made landscapes lack distinctiveness and have little connection with their geographical contexts' (Relph, 1976: 21).

Placelessness is an attitude that is influenced by an overriding concern with efficiency as an end in itself (Sampson & Goodrich, 2009). For example, various technological advances such as mass communication and increased mobility, particularly tourism, erode the geographical conception of place distinctiveness. Therefore, significantly diverse places of the world are casually replaced with anonymous spaces, inviting discussions regarding place and placelessness, as the debate goes: 'place was good, placelessness was not, most things modern were eroding good old places' (Relph, 2016: 269). Eventually, the term place-making is utilised as an extended discussion of place and placelessness to describe the process through which a place derives its identity (Freestone & Liu, 2016; Lew, 2017; Lin & Lee, 2020).

The concept of place-making extends from public space to place in general, and from urban planning to sociology, politics and geography (Wise *et al.*, 2021). Not only place experts, such as architects and planners, but 'all of us as human beings transform the places in which we find ourselves into places in which we live' (Schneekloth & Shibley, 1995: 1). Place-making is 'the set of social, political and material processes by which people interactively create and recreate the experienced geographies in which they live' (Pierce *et al.*, 2011: 54). Place-making studies discover how places come into being (Gieryn, 2000; Lew, 2017). The root word (i.e. make) highlights the role of humans in shaping place identities regarding the locale and the meaning. The use of the present form of the verb + ing aims to emphasise that place is always in a state of becoming: it is a process and in progress (Crang & Thrift, 2000).

Sense of place

SOP can be interpreted as belongingness to a place, describing the human–place relationship. Research on SOP is an important component of 'place theory' within the broad discipline of social sciences, which focuses on experiences, such as personal encounters in a place (Jorgensen & Stedman, 2001, 2006). SOP comprises not only one's perception but also one's feeling about a place (Castree, 2009). Such personal feeling refers to 'the emotive bonds and attachments that people develop or experience in particular environments' (Foote & Azaryahu, 2009: 96). However, SOP is a vague concept due to its multidimensional and multidisciplinary construction (Relph, 2001). It is subsumed into various place-specific constructs, such as place attachment (Williams & Vaske, 2003), place identity (Nielsen-Pincus *et al.*, 2010), place dependence

(Kyle *et al.*, 2004), place belongingness (Hammitt *et al.*, 2006) and social bonding (Lee *et al.*, 2012).

The SOP measurement scale is broken down into several multidimensional constructs to widen its empirical spectrum, of which all underlying dimensions are regarded as an 'umbrella concept' (Kianicka *et al.*, 2006). All subjective and affective qualities towards the feelings of a place are interrelated attributes under the SOP concept. Jorgensen and Stedman (2001) combine place-based concepts (i.e. place attachment, identity and dependence) into a single SOP measurement model. They propose a multidimensional SOP scale that comprises: (i) place identity: beliefs about the relationship between self and place (cognitive dimension); (ii) place attachment: feelings towards the place (affective or emotional dimension); and (iii) place dependence: exclusive behaviour of the place compared with alternatives (behavioural or conative dimension). The scale is validated as an effective tool for expressing the thoughts, feelings and behavioural commitments of respondents towards a physical setting. Those dimensions are proven to be applicable to other non-physical environments to assess the attitude of people towards various spatial settings, including place identity and attachment, to examine the attitudes of local residents towards the impacts of tourism development. The results of the related studies reinforce the intertwined role of people's emotional bonds with a place (Strzelecka *et al.*, 2017) and arguably, the unknown theoretical underpinning of the notion of place identity on relevant global studies in culture and power.

Given the global flow of people, information and commodities, considerations on the progressive concept of place necessitate a new agenda to connect the concept of place as well as the associated SOP and place identity to the wider world (Massey, 1994; Younge, 2019). Intimate individual experiences in a place are not limited to the structured social actions within a specific location, but 'manifest as any form of communications necessarily invokes linguistic and theoretical abstractions of self, other, site and world' (Larsen & Johnson, 2011: 641). An 'open sense of place' is harnessed for transformation, growth and compassion based on the encounters of flow, orientation and exchange in a global place. It may be a practical social occasion allowing the 'expression and awareness' of situatedness that engenders the possible social and ecological relationship (with other places and people). Therefore, a placeless experience (MacLeod, 2006) in a generic festival such as the HKWDF is worthy of in-depth investigations against a placeness event of a relatively fixed and original island celebration such as the CCBF.

Place distinctiveness

Defining the features that lead to the distinctiveness of a place and distinguishing its own identity from other places are challenging, as people's feelings towards the local environment and its communities are essential to formulating their SOP. Thus, place distinctiveness is not limited to prominent physical aspects since the meanings of a place can also be shaped by invisible aspects such as festivals. Scholars (Corsane *et al.*, 2009) suggest that people's strong sense of local pride in the place that they represent will be the key characteristic to express their unique local identity. Festival participation, for instance, is an occasion where dialogues between visitors and local people can materialise to improve the understanding of a unique SOP. Turok (2009) conceptualises city distinctiveness by proposing the sources of differential advantages on which cities can focus. These are firms and industries; skills, knowledge and occupations; built environment and amenities; and image and identity. This framework considers city identity to be a cluster of the most easily 'amenable to alteration' and 'intangible' features among all sources. Accordingly, city promoters may use a sporting event to launch cultural activities (i.e. festivals) to rebrand a city's positive image and raise awareness of it on the world map (Lai & Li, 2012; Lau & Li, 2015).

The subjective and subtle qualities of a city (i.e. an appealing lifestyle) can also be highlighted via dynamic festivals in order to uphold a sense of city distinctiveness (Lau & Li, 2019). Festivals enable cities to inform about their outstanding myths and prejudices, strengthen local civic pride and instil a shared sense of identity among the community residents. So, the role of festivals is no longer to present a simple and single identity of a place for its local tradition and custom. A festival also serves as an effective tool for strengthening the togetherness of a community (Rogers & Anastasiadou, 2011), cultivating pride among local residents (Lin & Lee, 2020) and producing meanings of a place and people (Richards & Colombo, 2017). Following this line of research, this chapter focuses on the possibility that festivals could make or remake people and places. The ultimate goal is to generate an understanding of the extent to which the perceived sociocultural benefits of festivals articulate the production of a SOP and its respective locality.

Research Methods

The existing research literature helps to establish an overarching theoretical perspective that regards place as a social construction process. So, the main argument of the debate in this chapter is that place is irrelevant in certain cases because place is a product of human experience. Accordingly, the key issue for investigation is to explore how and by what means the placemaking of both festival celebrations materialises.

Sampling

The target population of investigation comprised the local residents who had attended either event. The number of visitors to the CCBF, recorded by the local press, reached 40,000 in 2013 (Chow, 2014) and 55,000 in 2018 (Su, 2018). In the absence of any specific statistical record, no demographic datum of the participants was reported. By contrast, the number of visitors to the HKWDF was officially recorded. The total number of attendances during the four-day event in 2012 and 2013 reached 188,000 and 140,000, respectively (HKTB, 2013b, 2014). This number was expected to reach 140,000 during the 2018 event (Tsui, 2018). On-site interview was the desirable mode of survey to reach the intended target samples because the uncertain demographic characteristics of both festivals could cause practical difficulties in securing a reliable sampling. Convenient sampling was applied to reflect the diverse views of festival attendance, to select the 'sample simply by including people who are available or can conveniently be recruited' (Hibberts *et al.*, 2012: 66).

Measurement

The main measurement constructs included the meanings of festivals, SOP and perceived place distinctiveness. In order to address the challenge that the constructs might not have been well established in the existing literature, the measurement items of this investigation were developed on the basis of interviews with local residents, and adaptation from the relevant literature (Figure 8.1). Other well-established variables were incorporated, including the SOP construct, festival meanings and place distinctiveness. SOP is regarded as a multidimensional construct (Jorgensen & Stedman,

Key constructs	*Measurement statements and empirical sources of references*	*Techniques used in this study*
Meanings of festivals	Developed by the authors	• Literature review • In-depth interviews • Panel review
<u>Sense of Place (SoP)</u> • Place attachment	Jorgensen & Stedman (2001, 2006)	• Literature review • Panel review
• Place identity	Hammitt et al (2006); Kyle et al (2003); Kyle et al (2004); Lee et al (2012); White et al (2008); Williams & Vaske (2003)	• Literature review • Panel review
• Place dependence	Hammitt et al (2006); Kyle et al (2003); Kyle et al (2004); White et al (2008); Williams & Vaske (2003)	• Literature review • Panel review
Place Distinctiveness	Developed by the authors	• Literature review • In-depth interviews • Panel review

Figure 8.1 Theory-technique roadmap

2001). It comprises place attachment (affective or emotional dimension), place identity (cognitive dimension) and place dependence (behavioural dimension). Based on the established measurement scales (Hammitt *et al.*, 2006; Kyle *et al.*, 2004; Lee *et al.*, 2012; Williams & Vaske, 2003), SOP was operationalised in three-dimensional constructs for this survey – place attachment, place identity and place dependence.

In the actual technique design, the SOP scale items were modified with the wording, particularly in relation to festivals. Considering the short duration of both festivals and to avoid a low response rate, the empirical study contained only four items under each dimension of the SOP construct. These measurement items were three-dimensionally operationalised in terms of place attachment, place identity and place dependence (Table 8.1). In addition, the distinctiveness of a place manifested three dimensions: production distinctiveness, consumption distinctiveness and identity distinctiveness. According to Markusen and Schrock (2006), identity distinctiveness relates to the extent to which cities are recognised by residents and non-residents as culturally unique. Tangible (e.g. built environment and amenities) and intangible (e.g. image and identity) elements are differentiated sources in the formation of a distinctive city. Initiatives proposed by local authorities aim to use 'the trigger of major sporting or cultural events to improve the perceptions of places' (Turok, 2009: 25). So, the feelings and perceptions expressed by the local residents were incorporated into the development of measurement items for the perceived place distinctiveness (Table 8.1).

The established measurement items were sent to a panel of experts for validity checking. They included three faculty members from local universities, two district councillors, two travel agents, two officers from the national tourism organisation and two local government officers. Two pilot tests were conducted to resolve validity and reliability issues. The first was run with a sample of undergraduate students (n = 100) in Hong Kong, who had attended the festivals, in order to further purify the items and scales. An exploratory factor analysis (EFA) was employed to determine the dimensions of the scales. Items with lower than 0.4 item-to-total correlations were eliminated. Cronbach's alpha was also performed. To further ensure content validity and reliability, a second pilot test was conducted, with another 100 questionnaires collected from respondents of both local and non-local festival participants. The data obtained were entered into SPSS and AMOS 19.0, and Cronbach's alpha with scores over 0.7 was considered acceptable (Richards & Munsters, 2010). An EFA with varimax rotation and a confirmatory factor analysis (CFA) were performed to verify the reliability of the factor loadings.

The individual standardised factor loading under each construct should be at least 0.50 and preferably 0.70 (Blunch, 2013; Byrne, 2010; Hair *et al.*, 2010b). Accordingly, construct validity was performed, and was assessed by both convergent and discriminant validity. Convergent

Table 8.1 Measurement items

Key Measurement Constructs	
Meanings of Festivals (MoF)	
1.	The festival presents local communal food and beverage.
2.	The festival includes diverse celebrating party elements (e.g. parade & performance).
3.	The atmosphere of the festival is full of celebrating party mood.
4.	This festival represents inherited tradition of the local community.
5.	The festival involves a high level of social interaction with local people.
6.	This festival presents a strong sense of community involvement.
7.	This festival contributes a sense of togetherness within the local community.
8.	This festival demonstrates social cohesion.
9.	This festival represents the contemporary living culture of the local community.
10.	This festival contributes to a sense of multi-culture landscape of Hong Kong.
11.	This festival presents a dynamic culture of Hong Kong.
12.	This festival presents a cultural icon of Hong Kong.
13.	This festival showcases how Hong Kong combines both local and global narratives.
14.	This festival enhances the community's self-recognition.
15.	This festival projects a sense of community pride.
16.	This festival provides a sense of continuity of the local community.
17.	This festival contributes to community well-being.
Sense of Place (SOP)	
1.	I feel relaxed when I attend this festival in Hong Kong.
2.	I feel happiest when I attend this festival in Hong Kong.
3.	Hong Kong is my favorite place to enjoy this type of festival.
4.	Everything about Hong Kong is a reflection of me.
5.	Joining this festival activity in Hong Kong says a lot about who I am.
6.	Taking part in this festival activity, I identify strongly with Hong Kong.
7.	Participating in this festival, no other place can compare to Hong Kong.
8.	I have a special connection to Hong Kong and its people.
9.	Hong Kong provides many opportunities to engage in my favorite festival activities.
10.	I enjoy festival activity along Hong Kong more than any other place.
11.	Hong Kong is the best place for what I like to do.
12.	I wouldn't substitute any other place for doing the types of things I do at Hong Kong.
Perceived Place Distinctiveness (Dist)	
1.	The festival projects a unique sense of local identity of Hong Kong
2.	The festival indicates a distinctive local way of life of Hong Kong
3.	The overall festival experience today is very unique
4.	I have not experienced a festival like this in any other place other than Hong Kong

validity is 'the extent to which it correlates highly with other methods designed to measure the same construct', while discriminant validity refers to 'the extent to which the measure is indeed novel and not simply a reflection of some other variable' (Churchill, 1979: 70). The objective of running both validity checks was to ensure that all those measurement scale items were unidimensional (Fornell & Larcker, 1981). Finally, a total of 17 items for the meanings of festivals, 12 statements for the SOP and 4 items for the perceived place distinctiveness were retained. Construct validity was assessed via convergent validity and discriminant validity to ensure the unidimensionality of all the measurement scale items.

Research Findings

The sociodemographic profiles of the respondents were fairly distributed in terms of gender. The majority were first-time visitors to the festivals, with 67% for the CCBF and 58% for the HKWDF. They were generally young to middle-aged adults with university degrees or a higher educational background. The student group shared 39.5% and 10.3% for the CCBF and the HKWDF, respectively. This may be attributed to the fact that the theme of the HKWDF is focused on the promotion of international wines and gourmet food, thus students may have less interest in such a theme. Over 80% of the respondents from the HKWDF were earning salaries of more than HK$30,000. By contrast, a considerable percentage of the sample groups from the CCBF were earning less than HK$30,000.

In terms of the measurement construct of SOP, an EFA was performed on the data to determine the underlying dimensions (Table 8.2). All factor loadings were above 0.5 with eigenvalues greater than 1. The internal reliability of each dimension was measured, representing satisfactory internal reliability. The CCBF and HKWDF respondents had a similar perception of the perceived items for SOP. Only one item – 'I enjoy festival activities in Hong Kong more than in any other place' – under the dimension – 'Place Dependence', statistically differed between the sample groups ($t = 4.45$, $p < 0.01$). Attention should be paid to the effects of festival-induced SOP on place-making. Heterogeneity may occur when planning and promoting a transplanted festival to the residents. The sociocultural meanings of a festival, such as historical and geographical contextual frameworks, are potential circumstantial factors that can articulate the production of a unique SOP.

Another place conception investigated was perceived 'place distinctiveness'. Markusen and Schrock's (2006) idea was adopted by

Table 8.2 Mean comparison of festival-induced SOP

Dimensions and items description	Festival A - CCBF (n=378)		Festival B - HKWDF (n=360)		t-value
	Mean	SD	Mean	SD	
Place Attachment (PA)					
1. I feel relaxed when I attend this festival in HK.	3.53	0.99	3.56	1.01	-0.42
2. I feel happiest when I attend this festival in HK.	3.65	0.88	3.56	0.95	1.25
3. Hong Kong is my favorite place to enjoy this type of festival.	3.59	0.91	3.56	0.99	0.47
4. Everything about HK is a reflection of me.	3.18	0.98	3.19	1.06	-0.16
Place Identity (PI)					
5. Joining this festival activity in HK says a lot about who I am.	3.35	0.92	3.36	0.92	-0.07
6. Taking part in this festival activity, I identify strongly with HK.	3.53	0.92	3.47	0.97	0.82
7. Participating in this festival, no other place can compare to HK.	3.10	1.09	3.22	1.14	-1.53
8. I have a special connection to HK and its people.	3.54	0.88	3.55	0.96	-0.11
Place Dependence (PD)					
9. HK provides many opportunities to engage in my favorite festival activities.	3.66	0.82	3.59	0.91	1.07
10. I enjoy festival activity in Hong Kong more than any other place.	3.43	0.90	3.13	0.94	4.45**
11. HK is the best place for what I like to do.	3.35	0.93	3.28	1.02	0.99
12. I would not substitute any other place for doing the types of things I do at Hong Kong.	3.13	0.95	3.05	0.99	1.04

utilising a four-item measurement scale to understand the perceived place distinctiveness. The findings satisfactorily exhibited a high degree of internal consistency. Hence, the construct that measured perceived place distinctiveness was valid (Table 8.3). The CCBF respondents perceived the meanings of a festival in the projection of a unique sense of local identity to be considerably higher than those for the HKWDF ($t = 12.50$, $p < 0.01$). This result is rational because the CCBF is a local community-initiated festival with a 100-year history. Concerning not only the notions of place in geographical discourse, but also the affective meanings towards a place contributed to community self-recognition and the establishment of community pride. Alternatively, the attendants of the HKWDF might merely perceive this transplanted festival as a social gathering for hedonistic entertainment. The meaning of the HKWDF was perceived to be more prone to an international business opportunity.

The findings indicate, however, that local respondents agreed more on the consolidated linkage between the HKWDF and the global cultural representation of Hong Kong ($t = -17.70$, $p < 0.01$). The HKWDF highlights the multicultural place character, projecting a contemporary identity of Hong Kong. Even if the theme is common, the subject is connected to the prominent place positing of 'Asia's World City' that has been deeply implanted in the minds of both residents and tourists. The HKWDF creates an imaginative place identity that differentiates Hong Kong from other Chinese cities. Given that Hong Kong has a unique historical backdrop of multicultural privileges, the HKWDF is a widely accepted form of global event. The theme of the HKWDF effectively highlights the 'multicultural character' of Hong Kong, thereby shaping an 'open sense of place'. Once a festival is widely endorsed by the community, it can effectively promote a unique SOP even if its theme is inauthentic.

Table 8.3 Mean comparison of festival-induced place distinctiveness

		Festival A - CCBF (n=378)		Festival B - HKWDF (n=360)		t-value
		Mean	SD	Mean	SD	
1.	This festival projects a unique sense of local identity of Hong Kong.	3.94	0.76	3.10	1.04	12.50**
2.	This festival reinforces the national cultural identity of Hong Kong.	3.26	1.07	2.94	1.10	4.02**
3.	This festival highlights the global cultural representation of Hong Kong.	2.67	1.12	3.95	0.83	-17.70**
4.	This festival indicates a distinctive local way of life of Hong Kong	3.97	0.76	3.98	0.75	-0.23

Noted: * Significance at $p < 0.05$; ** significance at $p < 0.01$.

Conclusion

Previous studies of festivals have predominantly emphasised the contribution to the tourism economy, and may have overlooked the implicit sociocultural meanings that enable festivals to be a potential tool for place-making. The existing studies of the tourism–place relationship, especially those repelling a scientific-positivist methodological paradigm, have tended to overemphasise the role of tourism in unmaking places. Tourism in reality makes places, while possessing a dialectic role of unmaking places. In contrast to the previous studies, the case analyses presented in this chapter regard a small island place as the centre of meaning, offering an alternative illustration of such to uncover the simultaneity of the process that tourism dialectically contributes to place evolution. The analyses illustrate how a local heritage festival – the CCBF, and a transplanted global event – the HKWDF, may share similar SOP despite their distinct contextual differences. However, dissimilarities are also evident in the identified scales of place meanings. The CCBF appears to have instilled a sense of the traditional heritage of Cheung Chau Island and upheld the local scale of place meanings.

The HKWDF expands the international, transnational and multicultural identity of Hong Kong by signifying its profile of what scholars consider a 'supranational dimension of place' (Saar & Palang, 2009). This festival has been queried with regard to its role in differentiating Hong Kong from other cities in China. The sociocultural meanings determined from the HKWDF may act, arguably, as a catalyst to legitimise an open SOP (Colombino, 2009), because it obviously sheds light on the sense of 'cosmopolitan Hong Kong' (Rohlen, 2000) – a unique cultural landscape that differs Hong Kong from China's mainland. Thus, this chapter demonstrates a case in point that 'cultural politics has emerged as a means of either maintaining or resisting domination…, and in ways that demanded fresh and dedicated analyses' (Schiller, 1994: 1). The findings presented in this chapter suggest that a meaningful multicultural SOP can be produced, as long as a festival belongs to one of the following dimensions of place meanings: supranational, national, local and individual (Saar & Palang, 2009).

SOP constructs, therefore, may become mediating factors for enhancing the perceived place distinctiveness from the residents' perspectives. Some scholars suggest that identities are constituted through multiple understandings and experiences of diversity (Oke *et al.*, 2018). Although the theme of the HKWDF is not uniquely local, the subject is connected to the prominent place positioning of 'gourmet Hong Kong'. This international, transnational and multicultural style of wine-and-dine carnival may create an imaginative place identity that makes Hong Kong a unique place. Given the multicultural background of Hong Kong, the HKWDF is an acceptable form of transplanted event as perceived by both residents and tourists. It reflects more than the creation of a dynamic tourist attraction by staging 'hybrid identities' of Hong Kong. No magic formula exists to create instant success for the promotion of festival tourism. Nonetheless, one of the key criteria to success may be the selection of the festival theme.

SOP is attached to residents' daily life, while tourists' SOP can be developed when they participate in the social and cultural activities of festivals. Local identity thus 'becomes an ornament, a public relations artefact designed to aid marketing' (Knox, 2005: 38). Identity politics as a term is therefore used to relate to any kind of political intervention that intentionally targets a group inside which people share common attributes (Younge, 2019). If a city's primary festival no longer performs an identity narrative to which the population can subscribe, the community may choose to transform the local festival through new activities – creating new festivals with which the community identifies (Magliocco, 2001). Few 'transplanted' celebrations such as the HKWDF can be found elsewhere. This chapter reaffirms the vision of de Jong and Dannecker (2018) and Waldinger (2013), namely increased awareness of the spatial and cultural interconnectedness of people, cultural forms and objects as well as economic processes will compel researchers to rethink their perspectives and agendas of studying identities, global culture and power.

9 Linguistic Landscape, Tourism and an Island Place Making

Introduction

In this chapter, the key issue for discussion involves the manifestation of the linguistic landscape on a small island, namely to what extent and with what characteristics place meaning 'translates' into linguistic landscape. The case study in point – Wailingding Island – is located off China's southern coast under the jurisdiction of Zhuhai City of Guangdong Province (Figure 9.1). Approximately 21 km from Lantau Island of Hong Kong, Wailingding Island is in the Pearl River Delta region that has played a vital role in the opening up of the Chinese economy since the 1980s, when a special economic zone in Zhuhai was established. The region will also likely continue to play an important role within Chinese economic and geopolitical endeavours as the government deepens regional economic integration with the Guangdong–Hong Kong–Macao Greater Bay Area Plan while extending the outward push under the Belt and Road Initiative. This latter policy not only strives to build new infrastructure projects to connect Asia and Europe, but it also seeks to encourage tourism along the ancient maritime Silk Road routes (China Tourism Academy, 2018).

Zhuhai governs numerous small islands of the Wanshan Archipelago lying to its southeast, well known as the 'City of a Hundred Islands'. One of the islands – Wailingding – is designated a national four-star tourist resort for its geographical location and sun, sand and sea resources. Although the COVID-19 crisis put China's international tourism on hold, domestic tourism still saw steady growth even under the strict measures that China's central state adopted to curb the pandemic. Within and/or cross-province travel, especially for holidays on a resort island such as Wailingding, became very popular through 2020–2021. Thus far, however, the small islands of Zhuhai have received little attention within tourism studies, with two key exceptions (Hong, 2017; Wang & Bennett, 2020). However, neither exception has taken the linguistic landscape as an approach to explore the multilingual phenomena prevalent in tourism. The case study presented in this chapter aims to fill this research

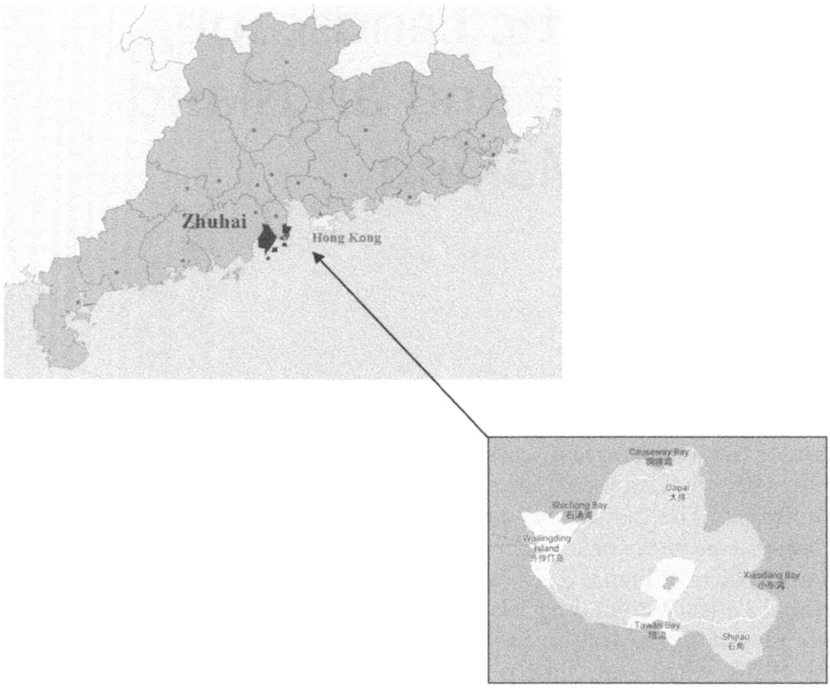

Figure 9.1 Location map of Wailingding Island (Adapted from Wikipedia, 2022)

gap. It focuses on this most visited island of the archipelago, to explore the question to what extent and with what characteristics place meaning 'translates' into linguistic landscapes.

Literature Review

Islands are popular tourist destinations, as pointed out in Chapter 8. However, studies of island tourism predominantly focus on planning and development strategies, policymaking, economic benefits and sustainability, marketing, as well as host–guest interactions. Previous studies of island tourism mostly adopted cost–benefit approaches, neglecting the human–place interaction. This chapter first establishes a dialogue between 'island' and other key concepts such as 'place', 'tourism' and 'linguistic landscape', in order to ground the specific research on a solid theoretical base.

Island and place

The key concept 'islandness' represents the 'signature of the island condition' which refers to the contradiction between the 'openness and closure' of islands, or their connectedness with and isolation from the mainland as well as other islands (Baldacchino, 2004). Scholars view the concept as traits and symbolic meanings that islands possess both in the real world and in the imaginative world (Baldacchino, 2008): those of remoteness, peripherality, tenacity and the physical surroundings of the seas (Hay, 2013). Scholars identify the traits and meanings from different perspectives: either a more binary one that categorises them into 'fault lines' (Hay, 2006) or a more relational one that mediates the contestations among the conceptions and focuses on islands as 'relational spaces', with relationalities such as land/sea, island/mainland and island/archipelago (Stratford *et al.*, 2011) as the research focus. This latter line of thinking also questions the static conceptions related to islandness and is regarded as being the basis for the 'relational turn' in island studies (Grydehøj, 2020).

Geographers consider island conditions through the lens of human experiences, believing that the metaphorical island should be attended to first. Typical images of tropical islands as 'Eden' and as 'remote to the continent' are used to reveal how people articulate their attitudes towards certain environments by focusing on landscapes that have acquired special symbolic significance (Tuan, 1967). To understand 'islandness', the lived experiences of people on islands, temporarily and permanently, must first be investigated. They include islanders, tourists, migrants and even nomads and drifters. Thus, the supposed inclusiveness in the humanistic perspective naturally invites a question for island studies, concerning positionality and responsibility – for whom are island studies undertaken? The above-mentioned perspectives should be considered because they inform how people construct their lived worlds, and, in this chapter's geographical context more specifically, islands as places where people live. As places, islands are both tangible entities and constructed places of experiences. 'The lives of islanders are inextricably bound up in island metaphors, symbols, and representations crafted in negotiation with the outside world' (Grydehøj, 2017: 6). A rigorous study of islands, therefore, needs to consider both the perceivable physical and metaphorical aspects of islands (Grydehøj, 2020), by scrutinising ideas of place in its sense of being an experiential notion.

The geographical idea of place includes three components: geographical location, material form, investment with meaning and value (Gieryn, 2000). Location refers to a certain spot in the universe, while locale is the sphere of action and interaction. Geographers probe people–place relationships through the realm of experience, arguing for the subjective experiences of people (Buttimer & Seamon, 1980; Relph, 1976; Tuan, 1977). Place is 'what takes place ceaselessly, what contributes to history in a specific context through the creation and utilisation of a physical setting' (Pred, 1984: 279). In a world of social hierarchies, place making is seen as a tool in creating, maintaining and transforming relations of domination, exploitation and oppression (Lau & Li, 2015, 2019; Li &

Lau, 2022). It is an innate human behaviour, ranging from the organic and unplanned actions of individuals – defined as 'place-making', to planned and intentional global theming by governments – defined as 'placemaking' (Lew, 2017). Placemaking may carry negative connotations of the notion of placelessness or 'non-place' – a temporary space for consumption and communication, such as shopping malls where the human role is that of customer (Augé, 1995; Kwong & Li, 2020). In today's world, places and spaces, places and non-places tangle together, indicating that conventional ideas about location, context and sense of place no longer apply since places and the way they are experienced are fundamentally impacted by technological and social changes (Relph, 2016).

Increasing human mobility imports ideas and practices from other places that are integrated with whatever is locally distinctive, for example, the host–guest interactions in a tourist destination. Places are thus regarded as social construction processes, conditional to both the level of globalisation and the way in which the processes are mediated through local place making. Bosman and Dredge (2011) propose that the best tourism places exhibit both place making and placemaking. Other scholars believe that 'local distinctiveness is not just a beautiful place; it is about details… and the things which create identity' (Corsane *et al.*, 2009: 50). Beyond the tangible aspects of a place, meanings can be shaped by intangible specifics such as an island festival (Giles *et al.*, 2019) or the linguistic landscapes (Lou, 2012). Tourism produces a unique site of intergroup discourse; it creates not just a location, but a public space of intergroup mutual place making – the 'physical co-presence' that establishes a respective community (Lau & Li, 2019; Li & Lau, 2022; Richards & Munsters, 2010).

Tourism and linguistic landscape

Tourism is founded on the appeal of distant places (Cao *et al.*, 2019; Xue & Zhang, 2020; Wang, 2000) where unique social-cultural and natural landscapes entice people to cross spatial expanses, thereby establishing the socialisation space for new forms of expression and social patterns of action (Graburn, 1989, 2008; Urry, 2006; Urry & Larsen, 2011). Scholars regard tourism as a discourse of difference through which power is propagated by the narratives of practice (Ashworth, 1994; Li & Hu, 2008; Li *et al.*, 2010; Li & Mai, 2022; Zhang *et al.*, 2015). Discourse constrains individual agency by shaping social relations and identities (Foucault, 1980), and naturalises and universalises people's worldviews (Thorpe, 2012). Due to its contribution to socialisation processes, tourism can play an important role in shaping values, behaviours and identities (Lau & Li, 2015, 2019; Morgan & Pritchard, 1998). Tourism therefore has broad social-cultural meanings that extend far beyond the consumption of places. Cheer (2020) finds that an attractive landscape provides the potential to mitigate the disadvantages of small islands' tourism regarding infrastructural provision. However, this requires establishing an effective development policy, resource management systems and marketing strategies that promote unique human and natural landscapes (Wang & Bennett, 2020; Weis *et al.*, 2021). Studies find that in linguistic landscapes some tendencies preconditioned by language attitudes and tourism do relate to people's needs and exchange (Ruzaitė, 2017).

Landry and Bourhis (1997: 23) define the linguistic landscape as 'visibility and salience of languages on public and commercial signs in a given territory'. Geographers consider language an important element in shaping place, having the power to create and maintain locality (Tuan, 2006, 2013). Geographers are concerned more about the symbolic function of the linguistic landscape that shapes the geographical space into a social space, triggering the production and consumption of new space. However, geographical inquiries of the linguistic landscape within China have mostly focused on the calligraphy landscape: debates over the relationship between calligraphy and the urban landscape, the spatial attribute of the Chinese calligraphy landscape and the implications of those for human–place relationship. They regard calligraphy landscapes as tourist attractions to discuss the effects on tourist sense of place (Xiao *et al.*, 2012a), image perception (Ke *et al.*, 2010) and residents' place identity (Xiao *et al.*, 2012b). International studies have revealed that through the infiltration of political forms into the linguistic landscape, local political struggle can be manifested in a language (Choksi, 2015); different federations maintain local differences due to decentralised linguistic policy (Lanza & Woldemariam, 2014); inconsistencies between national consciousness and the identity of the local residents in the linguistic landscape of a place (Lou, 2012). Language is thus an important medium for people to experience and understand the world. It is, after all, the carrier of cultural value and significance.

In sum, language is both a type of tourism resource and a part of tourist destination culture. The linguistic landscape accounts for one of the main contents of geographical inquiries on tourism and place. Language and place making are inseparable in the process of tourism development. Tourist activities help promote the vitality of a local language, enhance the uniqueness of the linguistic landscape which, in turn, offers tourists unique experiences (Kallen, 2009). An important characteristic of contemporary society is mobility, and locality is believed to be dynamic with processes of continuous making and remaking of place identities. The spatial movement accelerated by tourism generates various factors in time and space. The linguistic landscape of a tourist destination has a symbolic function which exists in places and manifests specific values, worthy of in-depth investigations.

Research Methods

The above theorisation of literature posits tourism as a discourse, offering an ideological prism through which to examine the realms of the sociocultural and spatial politics in the island's tourism and linguistic landscape. Accordingly, the extent to which those discourses are accepted and subverted by both host and guest communities, and the ways in which they shape the linguistic landscape are contemplated in reference to Structuration Theory (Giddens, 1984). Under the formulation of the tourist gaze (Urry & Larsen, 2011), the power of discourse is enacted at the social level, creating a dialectic between structure (officially planned tourism) and agency (its receipt at the societal level).

Agents often face intractable structures in the form of rules and social classifications in their day-to-day lives. The perception of a duality between structure and agency is heightened by the fact that human action is short term whereas social structures appear more embedded across time and space. This principle of research is thus adopted, namely the recursive relationship between the duality of structure and agency in which each component can be researched through empirical evidence (Figure 9.2). This is to interrogate more rigid dualisms between structure and agency. In other words, examining the state mechanisms of place and subjectivity creation on the one hand, and individual understanding of the self, society and place on the other. The rationale is to probe the extent to which tourism is a medium through which people create and

1: External social structures – ***conditions for action***

Mediated through '*position-practices*', a social position associated with identity and practice together with a network of social relations (institutions, power relations, technologies etc.).

2: Internal social structures – ***agents' capabilities & what they know of the world***

a) General dispositions: social-cultural schemas, discourses, world-views, moral and practical principles, attitudes, embodied skills, personal values and commitments. Many of those dispositions are not amenable to change within the individual

b) Conjuncturally specific knowledge: relative to the immediate strategic terrain and the expectations of behaviour and action in relation to it (based on understanding of external structures)

3: Active agency and action – ***agents draw on their internal structures in routine or strategic ways***

4: Outcomes – ***as they feed back onto position-practice network, internal and external structures are either reproduced or changed***

Figure 9.2 Quadripartite approach in structuration theory (Adapted from Stones, 2005)

recast individuals' own subjectivities through perceptions of the materiality (e.g. linguistic landscape) of a place.

Investigation techniques

The key issue for investigation involves the manifestation of the island's linguistic landscapes, from which emerges the question of how and why this island place is created. In the context of tourism, differences between agents' communities are not only institutionalised but also emphasised on a daily interactional level. Accordingly, three interrelated considerations are highlighted to formulate specific investigation techniques: (i) place – an account of the social production of space; (ii) society – an understanding of the ways in which agents interact to produce and reproduce social relations and places; (iii) power – outline of the specific language texts through which power is made inert in and enacted through social relations, forms of knowledge and places. Those considerations enable the contemplation of a fuller picture about how tourism is enacted from the nation-state level, received at society level and then how social action feeds back into the linguistic landscape. Hence, three investigation techniques are formulated. First, tourism policy and promotional materials are considered through a discourse analysis along with the bulk of the empirical investigation of the materiality that projects the island's linguistic landscape. Second, the empirical investigation is manifested as interpretive techniques centred on interviews. Third, observation of interactive participants is used to supplement those techniques.

Research Findings

Wailingding Island covers an area of 4.23 km^2 and is close to Hong Kong. The geographical location and the sun, sand and sea resources have enabled this island to become a national four-star tourist resort. Tourist arrivals to the island exceeded 250,000 in 2016. Although the global tourism industry suffered tremendous loss in 2020, tourist arrivals to Wailingding Island still recorded growth because the COVID pandemic diverted Chinese tourists' attention to domestic holiday resorts (Zhiyan Wisdom Group, 2021). A record-breaking 42,000 domestic tourists visited the islands of Wanshan Archipelago within the eight-day National Day 'Golden Week Holiday' of 2020, which is a year-on-year increase of 10.5% (Pengbai News, 2020). With tourism development comes outside influences to Wailingding, the most visited island of the Archipelago. Those influences are inevitably changing the linguistic landscape of this former 'provincial backwater'.

Government dominance

The empirical investigation on the island is, by large, visual inspection, considering its small scale. All elements are counted as valid samples, including the linguistic landscape appearing outdoors, excluding

Table 9.1 Number of samples inspected as of November 2020

Types	Inside Tourist Area	%	Outside Tourist Area	%	Total	%
Monolingual	198	44.8	104	54.2	302	47.6
Multilingual	244	55.2	88	45.8	332	52.4
Total	*442*	*100*	*192*	*100*	*634*	*100*

those without texts or with unclear texts from the statistical scope. When a sample has both front and back sides or multiple flanks, each side is counted. All the inspected samples are recorded in Table 9.1. Among the samples, the linguistic landscape elements officially created by the government account for the largest proportion. The government constructs them mainly for the macro purpose of developing a popular tropical resort island through designing and producing, for example, the road signs, tour path directory boards, scenic spot highlights and directories, as well as the Chinese calligraphy arts display posters. The island administration started to launch tourism projects in the 1980s. After Hong Kong was handed over to China in 1997, the development of tourism was accelerated due to the relatively relaxed economic policy offered by the central government at that time. For instance, direct ferry services from Hong Kong were allowed and gradually drew the attention of Hong Kong holidaymakers to the island, increasing tourism investment. After two decades of effort focused on designing and planning, a relatively well-established tourism business emerged, and eventually in May 2017, the island was designated a national four-star tourist resort.

According to *Quality Grade Evaluation and Division of Scenic Spots* (China National Tourism Administration [CNTA], 2015), the languages in panoramic logos or panoramic guide maps for tourists must be in Chinese, English or other official foreign languages. Signboards, scenery introduction boards and guide maps must be coordinated with the environment of a tourist site. Since 2016, Wailingding Island Administration has invested a substantial amount of capital to build a high-end brand of local tourism, through comprehensively improving the identification system of the scenic sites. For example, a two-dimensional code is added to all the scenic site introduction boards. The new designs present the local history with the island's distinct features of both human and natural environments. Every indication shows that the government dominates the design and construction process, defining the overall style of the linguistic landscape through the implementation of tourism policies and other administrative regulations.

Samples of the government unified *Samples of various private businesses*

Figure 9.3 Samples of linguistic landscape on Wailingding Island (Source: Author)

Private sector involvement

Operators of the private sector also contribute to constructing the linguistic landscape (Figure 9.3), and they care more for tourists' needs. However, most private businesses care less about the official language standards. The standards require that English is used to facilitate tourist communication. However, private business operators who run mostly small ventures do not possess language skills other than Chinese. So only a few, such as tea houses, cafes, bookstores and luxury hotels, present business bilingually – usually in Chinese and English. They are a younger group with higher educational backgrounds. They believe that adding English sign translations suits the image of their businesses and will help attract consumers. A few private shops use traditional Chinese characters commonly seen in Hong Kong, Macau and Taiwan. They consider the traditional Chinese characters an inheritance of the past, suitable for local identification.

Traditional catering businesses, local time-honoured stores and modern-style new shops are the main contributors from the private sector. The signboards of the catering businesses are simple, mostly made from plastic or iron frames, and are in simplified Chinese characters. The local time-honoured stores, whose businesses mainly cover daily necessities such as fishing gear or groceries, like to present their business in traditional Chinese characters. The modern-style new shops are run by relatively young owners with short business histories on the island. They pursue fashionable and personalised designs, such as neon lights and electronic displays, presenting their businesses bilingually in Chinese and English. They make their signboards individually, with free choice of design and presentation, as revealed by Shop Owner 1:

> '哈味 [Ha Wei]' comes from my wife's surname '魏 [Wei]' that is homophony with the Chinese character '味 [wei]', referring to 'flavour and taste'. The essence of running catering business is 'flavour and taste', so, giving my shop the name '哈味[Ha Wei]' is to make it appear down-to-earth, approachable and friendly. Besides, '哈 [ha]' has the edge of the Chinese character '口 [kou]', referring to 'mouth', it symbolises 'eating'. (Direct interview quotation from field research)

Shops affiliated with government institutions also have their own brand design of signboards, displayed in a unified chain form. They present individual characteristics in line with typical island features, as revealed by Shop Owner 2:

> We have unified signboards to indicate that we are of chain businesses, but each shop designs the posters out of individual characteristics. Look, that vertical bar shape on our signboard symbolises sides of books. As for other posters to be displayed outside, they certainly vary in terms of locations of shops. Say, our shop is on an island, no doubt we present business with island holiday characteristics. (Direct interview quotation from field research)

The investigation shows that whatever the rationale behind the designs of the local business presentations, the operators in one way or another point to the factor of enticing tourist consumption. As a national four-star tourist resort, stimulating consumers' interest in taking holidays on the island accounts for the fundamental principle of presenting the local businesses.

Tourist perception

The findings show that it is not in the tourists' interest to appreciate the particularities of Wailingding Island, such as the local culture and customs. The island is, after all, a holiday playground on the tourists' doorstep, especially when the COVID-19 epidemic made overseas holidays impossible. Tourist 1 states:

> For us mainlanders visiting an island and seeing the great ocean are really something! You may come to the seaside watching the sunrise in the morning, enjoying the peace and tranquility that are so rare in the roaring urban expansion of the mainland cities. Well, luckily China is huge that we still can travel within our own country, although overseas holidays are our of the question now due to COVID-19. (Direct interview quotation from field research)

The linguistic landscape does play a vital role in satisfying tourists' needs because it serves as the main guide and the symbol that channel and stimulate tourists' visual consumption. No matter whether the signs and commercial posters are unclear and unattractive, they influence tourists' consumption preferences. First, the linguistic landscape related to unified official design appears reasonable and clear in content and quantity and effectively informative. For newcomers visiting the island, the unified welcoming posters, danger warning signs and road maps enhance their sightseeing convenience. Tourist 2 points out:

> I have noticed the road signs. I think they are well done; they provide us with useful information. I look at them and see they are well matching the scenery too: good integration of both human and natural elements. These road signs on this cute stone came to my eyes right after I got off the ferry yesterday, making me feel so good. (Direct interview quotation from field research)

Second, the officially unified linguistic landscape relates to both tourism and local community life. Conversely, private contributions to the linguistic landscape are full of life and vitality which is also manifested with the commercial atmosphere, similar to China's mainland urban areas. Some tourists thus complain that private businesses such as restaurants and drug stores have only simple and ordinary posters and signboards. They feel disappointed to see the island changing in similar ways to the modern cities on the mainland.

Third, tourists affirm that an interesting and novel linguistic landscape showing the island culture and life will stimulate their curiosity to explore more, as told by Tourist 3:

> I think diversified private businesses encourage competition, and as the result they will attempt to present their different features. These will, of course, benefit tourists. When tourists travel, they want new and different experiences. Say, when a store or a restaurant appears special and different to all the others, certainly I will be interested in having a look inside it. This is the first step of me to spend and buy in a place. (Direct interview quotation from field research)

Fourth, tourists' opinions of the appearance of English in the island's linguistic landscape vary. Some think it is enough to use standard Chinese because visitors to the island are mostly Chinese domestic tourists. Signboards in a tourist destination are, after all, to inform and guide visitors. However, many others agree that using English on the island is necessary and feasible, since the English language indicates the island's global reach and modernisation. Tourist 4 points out:

> I think using both Chinese and English can be more effective for information transmission, because the island's visitors are from different places.

> There are mainland Chinese tourists here, there are also visitors from Hong Kong, Macau, and Taiwan who are more used to reading traditional Chinese characters, and even English. Their travel needs should also be taken care of. (Direct interview quotation from field research)

In a nutshell, the findings suggest that the officially defined linguistic landscape plays an important role in information transmission, manifested in the scenic spot panoramic maps, road signs, scenic spot signs, health check points as well as risk and danger warning signs. In the meantime, the privately designed linguistic landscape contributes, more symbolically, to maintaining the atmosphere of the local culture and customs. In the tourist gaze, however, the island's visitors as holidaymakers usually pay more attention to the overall visual effects of the landscape than its contents.

Conclusion

Based on empirical research, this chapter discusses how tourism, as an important factor, may influence the construction of the linguistic landscape of an island resort in China. Such influence is manifested, specifically, in government-planned development and the local private sectors' involvement, in both the material contents and the symbolic embodiment of the landscape. In the tourist gaze, the main consumers – tourists' preferences for certain elements of the linguistic landscape – can provide useful insights for shaping the linguistic landscape of a given place. The basic characteristics of the landscape are identifiable from the perspectives of the government, the private sector and the tourists; whereas the information function informs about the material morphological characteristics, as well as the purposes and principles of the main producers that create the landscape. It is essential that both the materiality and the inherent cultural values of the linguistic landscape of a given place be examined. Thus, this chapter affirms that the place making effects of the linguistic landscape, in the case of tourism development on Wailingding Island, are evident in the following aspects.

Informing of local narratives

The island's local narratives, in both folk customs and history accounts, are presented through the linguistic landscape. Unique celebrations such as the 'Fishery Re-opening Festival' is held annually, and its popularity is increasingly enhanced along with tourism development. Additionally, Chinese calligraphy arts are presented as part of local history, in the linguistic landscape, including the 'Cliff Stone Carvings' (Figure 9.4). The so-called 'Lingding Yang' (waters surrounding Wailingding Island) has served as the checkpoint fortress since ancient times. Relics of military facilities and air raid shelters from war times remain on the

Figure 9.4 Chinese calligraphy arts presented on 'Cliff Stone Carvings' (Source: Author)

island. The calligraphy carved on a stone is a copy of the late president of the People's Republic of China, Chairman Mao's handwriting, telling the story of an ancient hero whom the islanders feel proud of and are enthusiastic to share the related historical events with visitors.

Presenting cultural values

Place making effects are also manifested in the presentation of local cultural values, in both the institutional and the livelihood cultural aspects. Of the island tourism development related to the linguistic landscape are relevant national policies such as standards and norms of using languages at scenic sites. As a national four-star tourist resort, it is required that scenic site introduction boards are in accordance with nationally unified standards (Figure 9.5). The local administration also has its own 'Rules for the Implementation of the Regulations on the Administration of Outdoor Advertising Facilities and Signboards in Zhuhai Special Economic Zone' (Zhuhai City Government, 2017). Specific regulations are set for the private sectors to follow in terms of the location, size and language standards of their outdoor advertisement facilities and signboards. To maintain the quality standards of a national four-start tourist resort and coordinate with the natural environment, the scenic elements that contribute to making the island landscape are unified in blue and white colours no matter if they are from the government or the local businesses. The colours echo the blue ocean, sky and white clouds, in every representation symbolising the island's fishery livelihood culture.

Figure 9.5 Officially administered bilingual road signs (Source: Author)

Embodying local identification

The symbolic meanings of the linguistic landscape are constructed on social-cultural and economic factors, reflecting the island's social-cultural change. They are fluid and vary with the perceptions and understandings of different agents. A major force that shapes the linguistic landscape is tourism information transmission and communication, and multiple agents from the government, the private sector and the tourist group perceive and understand the symbolic meanings of their respective subjectivities which define and influence Wailingding's tourism industry.

The tourists' sense of place, namely how they understand and feel about the island, is influenced by the symbolic 'islandness': the landscape signs in blue and white colours. Interviews with diverse demographic respondents indicate that tourists' interest in Wailingding Island concentrates on the cliché of a 3-S holiday: enjoying the sun, sand and sea that a tropical island offers. When asked their feelings about the place, tourists frequently mentioned key words including pure nature, life simplicity, good sea food, relaxation, peace and tranquility. In this vein, Wailingding symbolises a makeshift holiday paradise against the global COVID-19 crises.

Migrants from the mainland account for the majority of the private business operators, who came at various times and have settled on the island for years. Their understanding and interpretation of the linguistic landscape vary, but more or less point to functional factors such as information transmission, being unique and attractive to stimulate consumption. They have witnessed the island's change in the process of tourism development. For them, the changing linguistic landscape symbolises hope, dependence and pride in this very place that they regard a 'second home'.

The government plays the dominant role in shaping the island's linguistic landscape by setting relevant regulations and standards, including the languages and sizes of signboards and commercial posters, the introduction of scenic sites and panoramic maps and the supervision of national values promotion and COVID-19 control measures on official bulletin boards. The government sees to it that tourism-related landscape elements are in accordance with those unified national rules and regulations, symbolising the nation-state control of the island. This infiltrates the national consciousness with the island's linguistic landscape, connecting this small, remote island with China's mainland, and assuring the islanders' security instead of isolated uncertainty.

To conclude, this chapter affirms that the linguistic landscape plays multidimensional roles in the place making process of tourism development. As the main representation and carrier of the island's history, heritage and culture values, the linguistic landscape with the perception, interpretation and understanding of the different agents manifests deeper meanings of the island beyond the displayed content. The texts are fundamental to making a meaningful place, but it is the cultural connotation that appears closely related to the social environment, manifesting deeper meanings of the island beyond the displayed content. Therefore, one could argue that the symbolic meaning of the linguistic landscape is fluid, including not only the subjective consciousness of the creative subject, but also the cognition, perception and understanding of the linguistic landscape by agent groups of various interests.

10 Tourism and Social-Cultural Change in China

Introduction

China's recent history of regional differences and social turbulence consequently split the culture and economy between the cities and the countryside. In the late 19th century, the 'New Culture Movement' introduced Western social evolutionism, positioning China's growing cities as the centre of the nation's modern civilisation, in contrast with the 'tradition' and 'backwardness' of the countryside (Ren, 2010). In the early days of the founding of the People's Republic of China (1949–1976), the countryside was associated with the feudal past, although being simultaneously regarded as the cradle of the communist revolution. At the same time, urban bodies were identified with modernity and progress, as institutional change gradually closed the door to cities for rural residents (Wang *et al.*, 2017). From this point of departure, this chapter sorts out China's rural–urban separation through an in-depth analysis of urban tourists' perception of rural places.

This chapter focuses on urban tourists' imaginings and performative acts, specifically, of their travelling through the countryside of the Southeast Miao and Dong Ethnic Minority Autonomous Prefecture in Guizhou Province of China (Figure 10.1). The lure of this place for tourists includes its natural landscape, rural scenery and ethnic minority cultures (Finkelman & Centeno, 2020). Due to its geographical remoteness, this place has been perceived by city residents as a true example of the key social-cultural and economic distances that typify the rural–urban separation in China's modern history. This chapter delineates the ways in which the countryside, as the 'other' place to China's urban centres, mediated by tourism, becomes transcended, rendering sublime the globalised city lives.

Literature Review

Perception of and interaction with places reflect the ways people understand the world in which they live, and help people determine what is, or is not, important to them (Cresswell, 2007). As profound centres

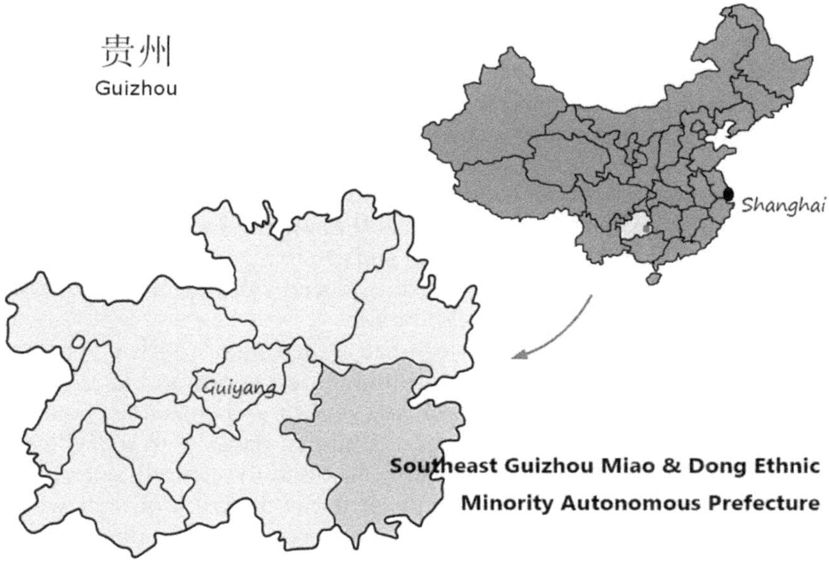

Figure 10.1 Location map of the case site (Adapted from Wikipedia, 2020)

of human existence, places are keys to connecting all human feelings to the material environment (Relph, 1976; Tuan, 1975, 2001a, 2001b). Places are significant because 'they are the focus of personal feelings… [and those] feelings given to places may be so strong that they become a central part of the identity of people experiencing them' (Rose, 1995: 88). Places, however, can be changed by forces beyond human control, raising a series of disturbing questions as to what happens when places change.

Harvey (1996) investigates those questions and points out that the advance in communicative technologies has, in many ways, made people feel that their places are under threat. He uses the term 'time-space compression' (Harvey, 1989) to describe the central features of the postmodern society where the rigidities of the economic systems are succeeded by the ever-expanding forces of globalisation. Similarly, Giddens (1990) uses the conception of 'time-space distanciation' to illustrate how facets of social reality and interaction are stretched across time and space. Individuals' social-cultural realities are thus increasingly shaped and reshaped beyond their immediate physical location in space and time. Both scholars predict that the advance in communicative technologies is to connect disparate places across geographical distances, thereby interweaving human societies.

Accordingly, other scholars put forward a so-called 'global sense of place' as a new theoretical framework for understanding places as processes, instead of sites of isolation and localism (Massey & Jess, 1995).

For instance, the emerging global cities along China's coastal regions manifest compelling explanations of the processes that this new framework is conducive to change; and demonstrates that outside forces are, eventually, impulses to locality reorganisation, challenging the theoretical hegemony of isolated places. However, this framework does little to address people's feelings of uncertainty brought about by globalisation, nor does it consider people's shifting attitudes towards changes in places. Consequently, a key issue remains: this framework does not help tackle the impending threat to both personal and group place identities enforced by constant alterations to the physical, social-cultural and emotional landscape of places that people call home.

Place is at the heart of tourism because the role of tourism is to display places that consist of people, communities, goods and landscapes, and to make them attractive for tourists to visit and consume. Tourism scholars should direct enquiries towards human nature – to study 'man-in-the-world', because a human being has a fully centred self and is able to transcend any given environment in the direction of their world (Tuan, 1965). The humanistic perspective sees three standpoints that exist as people's positionality: (a) as a member of the biological species of human, (b) as a member of a sociocultural group, and (c) as an individual (Tuan, 1967). From each of these standpoints, different dimensions of a human worldview may be considered, including the human body (the species), gender and cultures (the sociocultural group) and personal experiences and history (the individual). Thus, humanistic enquiry of tourism phenomena not only places an emphasis on people's experiences, but also places a more abstract focus on their ability to understand the world, namely the human consciousness. More than being just a mere location, place has meanings that are both symbolic and experiential. Those meanings can be uncovered by exploring the perspectives of the people who 'create' them through tourist activities (Lau & Li, 2015, 2019; Li, 2000).

In seeking to understand how people perceive the world, humanists value people's experience and self-consciousness. This chapter thus explores tourist understanding of rural China as the lived environment of people – from a humanistic perspective and by relying largely on the notion of place as the centre of meanings. As an industry and a practice, tourism is multifaceted by nature (Mostafanezhad & Swain, 2019). Some scholars had already raised a critical issue that multi-fronted research alternatives were highly needed for tourism research within China (Hollinshead & Hou, 2012) because the existing analyses tended to concentrate on the mechanism and content of state policy and management (Bao *et al.*, 2014). This issue remains unresolved, according to the author's review of the latest relevant documents (Academy of China Social Science, 2022; National Natural Science Foundation of China, 2022). The study presented in this chapter, therefore, can be regarded as an attempt to unsettle the deterministic approaches that dominate the academic sphere of tourism research related to China.

Research Methods

The investigation focuses on two types of urban populations: older nostalgic tourists (after 50s'–60s' generation) and young idealised travellers (after 80s'–90s' generation). The former can be considered one of the best-placed generations to contemplate China's changing social-cultural, economic and physical landscapes since the post-1978 economic reform. The latter, as the direct descendants of the so-called after 50s'–60s' generation, is best placed to feed from the former's nostalgia, thereby reviewing the constantly changing urban landscapes and reflecting on the transformation of China's rural–urban relationship. Broadly, two fieldwork techniques were determined: face-to-face interview and participant observation. Unstructured interviews were largely employed to give respondents the opportunity to develop answers naturally, because interviewing is to understand how individuals make sense of their own lives (Burgess, 1990; Li, 2000; Sullivan & Forrester, 2019). The interview topics included the impacts of tourism, tourists' perspectives and some wider issues of local, ethnic minority and Chinese culture.

Reflexive ethnography was adopted to observe and delineate urban tourists' perceptions of rural life that they encountered. The research fieldwork design was adapted, predominantly, to the possible situations of everyday life (Burawoy, 2003; Davis, 2007). In doing so, the researcher would meticulously observe both conscious and unconscious behaviours and discourses emerging in the process of interpersonal communications, thereby interpreting culture and society with in-depth analyses. The field investigations were concentrated over a three-year period between 2013 and 2016. The initial contacts with the investigation sites and the surrounding countryside began as far back as spring 2003. Guizhou Province changes with each passing year under the national poverty alleviation plan. To observe its latest changes, a couple of more brief visits were made in autumn 2017 and 2018.

Interpreting culture and society

The timing of the investigation field trips was chosen to align with specific peak tourist seasons. Spring is popular with photography enthusiasts as it coincides with the rice planting season, and summer is the traditional peak season for tourism. Both seasons were designated as the main time for fieldwork. More traditional 'deep immersion ethnography' and short-term 'rapid participatory assessment' methods (Chambers, 1991; Geertz, 1973; Li, 2000) were employed to understand the interaction mode between the residents and the tourists, and observe how tourism penetrates and impacts the place of investigation.

Morishita (2001: 10) suggests, 'social activities can be read for meanings just as spoken and written words are commonly read'. Deep immersion ethnography regards the researcher as the vessel for interpreting an assemblage of texts, thus enabling the researcher 'to be contextually sensitive, inductive and naturalistic' (Patton, 1982: 10) towards the phenomena in question. This approach also expects the researcher to become a key influence in the creation of knowledge through interpreting the words and actions of others (Jamal & Hollinshead, 2001). However, concerns may arise that some social and cultural interpretations that a researcher makes are beyond the realm of cognitive consideration for most laypeople. To reduce such concerns, non-representational field survey techniques, for instance, participant observation, are recommended.

Cultural interpretation may be understood, arguably, as a meaningful social activity, and its key is to be sensitive to the context, thus summarising it (Morishita, 2001). Human consciousness shapes cultural practice in both discursive and non-discursive ways. Ethnography field investigation, therefore, requires a researcher to capture non-discursive information (Latham, 2003). In general, social practice manifests (a) discourse awareness that explains what individuals do; (b) practice consciousness that determines beliefs and knowledge of oneself and others; and (c) unconscious thoughts that drive the 'natural' behaviour of 'instinct' or 'common sense'. Accordingly, a researcher is supposed to let the respondents tell their personal feelings naturally so as to interpret both conscious discourse and practice, while observing the feelings that the respondents may not be able to express for various reasons (Giddens, 1984; Larson, 2009).

The process of actual field investigation was thus manifested: (a) firstly, the investigator would invite the respondents to choose interview topics, and the respondents would often choose some daily life topics that did not need to be deeply thought about; (b) secondly, the investigator would ask them questions not related to daily life that needed recognition, because their interpretation of a personal experience could serve as a frame of reference to help analyse their inner world (Jamal & Hollinshead, 2001); and (c) finally, when the respondents had difficulty clearly explaining the reasons for some of their behaviours, the investigator would use observation to visualise their behaviours, thereby inferring what kind of experiences these behaviours might have accumulated. All human behaviours are intuitive. An investigator must go beyond the discursive to observe daily life events, in order to obtain the reasoning results behind the actual behaviours (Bernard, 2006; Morishita, 2001). The method of ethnographic investigation and interpretation not only enables an investigator to reflect, formulate and respect the creativity of social practice, but also provides a critical explanation for such practice.

Research Findings

If China's developed coastal regions manifest landscapes of connectivity between the central urban agglomeration and the world, then rural Guizhou – nestled in China's social-cultural and economic periphery – displays a sharp contrast. Its rural vista dominated by farming, the remote geographical location and the Karst ravines create the local collective identity of unity between human and nature. Urban tourists from afar see rural Guizhou as a 'place' of tradition – a timeless, authentic place where they may still find the 'real' China.

A place under the threat of change

Guizhou is often imagined as a geographically remote, poverty ridden and ethnic minorities inhabited place. It is this imaginary that contributes to the local tourism appeal. In Southeast Guizhou Miao and Dong Ethnic Minorities Autonomous Prefecture where this study was focused, ethnic minority populations account for 80.3% of the registered residents (Paihangbang [Ranking list], 2017). Tourism to its most well-known tourist attraction – Jiabang Farm Terrace Scenic Area – did not start until 2010. Before that, the ethnic minority villages in the vicinity of the scenic area were not prominent among hundreds of thousands of minority villages in rural Guizhou. Although the construction of the infrastructure started quite late, the number of tourists added to the statistics had steadily increased when the author made an initial visit for preliminary contact with the investigation sites back in 2003. A few years later, the beautiful scenery of the farm terraces was presented as part of the background of the closing ceremony of the Guizhou Pavilion inside the 2011 Shanghai World Expo. This area immediately caught the attention of photography enthusiasts, and the scenic beauty gradually boosted rural tourism to the prefecture (Figure 10.2).

There is a yearning for a pastoral life and landscape as the increasing cost of modernity creates a structure of feeling – the so-called urban gaze (Urry & Larsen, 2011). Rural life is imagined to be authentic, and its more localised and more grounded traditional lifestyle is regarded as a symbol of historical continuity on various scales, thus providing urban tourists with a refuge from the problems of modernity. Geographical remoteness, intensified by the mountainous terrain and lack of extractable resources, are the facets of the countryside in rural Guizhou. Harvey (1996) points out that globalisation makes it harder for people to connect to places because their attachments to rapidly changing locales are becoming less meaningful. It is the tourism industry – as part of the globalisation process – that offers people of modern society the opportunity to seek out authentic places beyond the normality of a globalised life (MacCannell, 1976; Vidon, 2019). It is through the employment of

Figure 10.2 Guizhou rural landscape (Source: Author)

modern communicative technologies that rural Guizhou, imagined as the periphery of China, becomes noticed to tourists:

> Internet images posted by other tourists can help us see where newly developed attractions are. We went to Guangxi Province last year and met some people like us who love photography. We still keep in touch with them, and we can also introduce them some new scenic spots, and learn about new places
>
> ... (Direct interview quotation from field research)

Tourists feel that globalisation has not greatly affected rural Guizhou, while it has largely penetrated China's coastal cities. So, they imagine the countryside of Guizhou Province as a cultural anchor in time and space, against China's continually changing social-cultural and physical landscapes:

> In two or three years, it [rural Guizhou] won't be like it is now... I'm afraid the villagers won't be like this either. Now the local people are still simple and warm-hearted. If they want to treat you, that is because they really want to be nice to you, the relationship between people in other places is not like this.
>
> (Direct interview quotation from field research)

A familiar place that used to exist

The above interview quotations reflect the quintessential authentic place of rural Guizhou is perceived by tourists as 'yuanshengtai' – a Chinese discourse that is rich in meaning and has image characteristics contrasting with various ills of modern society (Wang, 2008). Implied by tourism, 'yuanshengtai' is indicative of the ways in which the countryside has become both a place of cultural meaning to urban tourists and a space to display contemporary anxieties over the perceived ebbing of traditional influences on modern life, rapid social-cultural change and environmental degradation.

Field Note No. 1

A year later, these changes were observed in the village: a tourist reception centre had been created, with its English translation completely mis-spelt (Figure 10.3), from which a short distance away a noisy retiree photography group were gathering in front of a farm-stay. The group cordially invited us to travel with them. As soon as the tour started, an embarrassing incident happened – the tour bus got stuck in the mud and could not move, until an excavator was called to pull it to safety. This incident, however, excited the group of tourists, and some said, 'this is exactly what we want to encounter here', because it proved that they had bravely experienced risks in the backcountry.

Figure 10.3 A tourist reception centre in a rural village (Source: Author)

Some older tourists explained that visiting rural Guizhou brought back their personal memories of the bygone era of the 'Cultural Revolution' (1966–1976). During that time, they lived for extended periods in the countryside when they were young. Rural tourism for those retiree photographers thus becomes a nexus for the elicitation of a certain kind of mediated nostalgia for authentic places. The majority of the older tourists spent their growing years in the countryside and directly experienced traditional rural life. They are of the former 'zhiqing' – the youth generation that was sent from their urban homes during the Maoist era of the 'Rustification Movement' that started in the 1950s and reached its peak during the 'Cultural Revolution' period. Their past life experiences manifest a strong relationship between personal memories of spending time in the countryside and the nation's past in general.

The older tourists hoped to capture some idealised community, such as the former 'People's Commune' or something similar, during their rural travel experiences. They paid more attention to the authenticity of the countryside as well as the honesty and directness of the rural residents:

> If the villagers offer you a drink, they have no other reason than just wanting to have a drink with you. We feel the rural people are more honest, friendly, and down to earth, less calculating than people in cities.
>
> (Direct interview quotation from field research)

The older tourists often talked to each other of the virtues of hospitality among people in the past – the qualities that they said were more common in the old days, and they felt they could still experience these qualities in rural Guizhou. When asked if they minded the discomfort of the relative underdevelopment in the countryside, many of them would quickly point out that they had experienced much worse. For them, rural Guizhou and its pre-developmental purity, link to their nostalgic yearning for past urban places that have disappeared under China's rapid development and social-cultural change:

> There is a better environment here. You see, children can go down the river to catch fish. When I was a child, I could go swimming in the river with my friends in Shanghai. But now Shanghai is polluted so much that there is no place for those activities.
>
> (Direct interview quotation from field research)

Seeing rural children playing in places that are considered safe and free from materialism not only arouses the nostalgia of the older tourists, but also reflects the intergenerational differences between their views and the norms of contemporary society. One of the key reasons that the relatively non-material environment of the countryside becomes a space for the older tourists to recall and share their personal experiences is that they could reflect on similar situations in their past lives by observing its relative poverty. This reflection provokes idealised images of freedom that is perceived to be lost in an urban setting where, through the material changes of development, perceived freedom and places of innocence are lost. Despite having bitter feelings when remembering times of economic hardship, the older tourists felt their past experiences were valorised and could be reconstructed as a benefit in hindsight:

> Living in the countryside as 'zhiqing' was difficult. We were young and naïve during those days, and thought it fun to go to the countryside with our friends, away from our families. But the life in rural China was hard at that time. The peasants were starving but we could not help them, we had to look after ourselves. But living through that time was good because that experience enabled us to cope with anything. Now we have everything and life is easy, but people often worry about their future. Why do we worry now? It is really a question to ponder.
>
> (Direct interview quotation from field research)

The older tourists not only expressed nostalgia but also reflected, critically, on the present. It appeared that their narratives and feelings largely extolled on the rural immateriality of life, in comparison with the obvious materiality of life in cities:

> You can see all these peasants are poor and they are all the same. They will help each other build a house or harvest the fields. You can see they are simple and warm-hearted. People did not have material possessions before, so they cared more about each other. The Chinese society was like that when we were young. Before, people were more important in society: a friend was just a friend then, but now making friends may be all about creating personal benefits.
>
> (Direct interview quotation from field research)

For the older tourists, revisiting the countryside acts as a mirror on the experiences of change in modern cities where society is seen as more selfish, more materialistic and offers less opportunities to learn from hardships while the advantages of the present are taken for granted. Many of the older tourists felt that the communal and selfless values prevalent in their youth could still be found in rural Guizhou. The word nostalgia – derived from nóstos (homecoming) and álgos (pain/ache) – is a longing for a home that no longer exists or has never existed, a sentiment of loss and displacement (Boym, 2001). In contemporary China, nostalgia is a state of mind that imbues globalised cultures, bred by the emergence of popular remembrance that has accelerated alongside the great leap forward to modernisation.

When nostalgia is manifested in a homogenising fashion and sweeps through groups or generations, it is symptomatic of a widespread response to a collective past that is forever lost. Travelling through the countryside enabled the older tourists to reconnect to a past and better place that they felt was lost in the waves of the nation's modernisation campaign. Rural tourism thus goes beyond their reminiscence of former hardships, as a means to physically reconnect with past experiences, selves and society. The place of countryside acts as a refuge from modernity and allows the older tourists to reassess their personal histories. The social discourse of rurality that saturates tourism thus posits the characteristics of 'yuanshengtai' places of rural Guizhou as bounded communities of traditional human relationships. The countryside is therefore deemed simple and straightforward, a part of the idealised community existing in space and time with a fixed value, culture, place and people.

A place of ideal meeting reality

University students represent another important group of informants. They arrive through volunteer travel, with the purpose of helping the people living in poverty-ridden rural Guizhou. Although their form of travel is different from the older tourists, their imagination, perception and feeling of the countryside suggest a continuity of impulses for idealised individual and social-cultural change. Those are similar impulses that manifested, back in the 1950s and the 1960s, as the youth generations' enthusiasm for gaining personal growth and development through experiencing China's rural hardship.

Field Note No. 2

The summer holidays had begun the day before but, surprisingly, the basketball court in a rundown building was found teeming with activity. This structure was donated as part of the so-called 'Hope Primary School Project' – a charity that focuses on giving educational facilities to impoverished rural children. Several efficient-looking young people were directing a group of rural pupils playing basketball. They were part of the university students who had just arrived the previous day. Across the playground, some girl students sat together in the kitchen, peeling a large pile of potatoes. They said that they had to do everything themselves in the village during the whole summer. The trip to rural Guizhou was organised by the students. They planned lessons that spanned various subjects. While working there they had to bring, buy and prepare their own provisions and equipment.

At one point, a girl broke down in tears. She was crying 'I can't take it'. She was upset because she had been sweating all day and had just found out that the only water source for showering was a pipe in a vacant lot at the back of the school. This was also a public area where people washed their clothes, dishes and vegetables. Other girls quickly gathered around her to lift her spirits, and it was not long before she had joined the cooking group in the kitchen. All the volunteers had to undergo a shocking degree of material shortages compared to their normal urban situations, and obviously some felt it difficult to adapt. Chatting with them, the researcher found their perceptions and narratives of the countryside filled with idealised camaraderie, the will to test their determination and the spirit of learning through challenges. When they put the individual in the collective life, their memory would arguably be a part of the memory of the society (Sutton, 2001), continuing the collective nostalgia of the older tourists.

'Collective nostalgia refers to that condition in which the symbolic objects are of a highly public, widely shared and familiar character' (Davis, 1979: 222). Those young students' perceptions and narratives of rural Guizhou drew upon a framework of collective memory that is similar to the older tourists, which, arguably, demonstrates the ways in which public nostalgia can influence the ideal and behaviour of individuals despite having no direct experiential references. Unlike the older tourists, those university students saw their rural experiences as ways out of uncertainties in cities. In a sense, the idealised students were accessing a kind of public nostalgia through which to form a critique of the current sociopolitical situation. They were, in a sense, accessing the idealism created during China's pre-reform era, when the sociopolitical system was imagined as capable of handling complex problems with simple solutions (Barmé, 2000). Those young people were probably best placed to learn from their parents' collective nostalgia of the traditional socialist China, as direct descendants of the after 50s'–60s' generation:

> My parents did not want me to come at first, but my father also thought it was a good thing to let me face some difficulties... Our parents faced hardship when they were young and many people my age feel that we should learn from similar experience in order to learn something new, for personal growth and development.
>
> (Direct interview quotation from field research)

The student's wish to benefit from experiencing rural hardship is a sentiment shared by older tourists:

> The young people do not need to come... there is no reason for them to experience the hardship as we did in the past, but when we older people see them acting like us before, we feel very proud.
>
> (Direct interview quotation from field research)

The young travellers inherit the former 'zhiqing' spirit of idealism, and put it into practice through a contemporary form of rusticated pilgrimage – joining a volunteer programme to help impoverished rural people. By so doing, they reaffirm the narratives of the older tourists. For the former, their legacy is consolidated because the actions of the younger generation rationalise it and it is further clarified in the globalisation era when materialist discourse prevails. The obvious connection between both is, namely visiting the countryside not only shapes and idealises the authenticity and simplicity of the 'place' and 'people' of rural Guizhou, but also provides a space for self-improvement and self-actualisation. It is those ideas that turn rural tourism into a space of social critique, against the perceived ebbing of traditional influences in the face of China's rapid social-cultural change.

Field Note No. 3

The young travellers who came to rural Guizhou were similar: all had a university education and many had secured good jobs. They saw their rural travel as not only a 'holiday', but also an opportunity to learn about their country and understand the problems that they felt characterised contemporary China. Although the young travellers' travel motivations were similar to those of the older tourists, their performative acts appeared more enthused with purpose to both understand the social-cultural and economic distances that characterise Chinese society and examine the associated material effects. A young backpacker disclosed:

> China is developing very fast, but the cost is tremendous. People are very poor in rural Guizhou because rich people will just look after themselves. Life here is so hard that some children must walk for hours, climbing the mountains to go to school. Look at the old person over there (pointing to an elderly outside a village clinic. She came all the way from there (pointing to the village across the valley), walking four or five hours to collect the medicine here. So, it is good for me to experience the rural hardship, because I can understand myself while being here to help them.
>
> (Direct interview quotation from field research)

University student volunteers had similar feelings:

> We want to come here to experience rural life and help the peasants. We have better resources and opportunities in cities than they do. This is the reason why they need this kind of thing (Hope Primary School Program). (Direct interview quotation from field research)

The narratives of the young people suggest a contemporary version of the utopian idealism of the 'zhiqing' generation: experiencing rural hardship to help those less fortunate. This idealism carries both nationalistic and individualistic undertones of self-improvement, sacrifice and self-actualisation. The young travellers' motives are characterised by the idea that travelling to rural Guizhou and facing hardships provide a path to self-improvement:

> I am happy to be here, although I knew there would be nothing here, and we would face some hardship... For people our age life is easy because we are all from single child families and we do not have to worry about money. When we study our parents will make us food and clean everything for us, we do not have to do anything. So, I think it is good to come here because I can learn to do something by myself.
>
> (Direct interview quotation from field research)

The ways in which those contemporary young travellers are nurturing narratives of self-improvement indicate the shifting of the geographical space for tourism and point to a kind of metaphorically personal transition. Compared to the older tourists who are directly connected with the past, those young travellers' expectations for rural China are even more idealised, since they do not have direct life experience of the nation's pre-reform era. Instead, their imagined, nostalgic rural is constructed, both socially and politically, by the state media, namely selfless behaviour of individuals is an integral part of the virtues of the Chinese nation (Li *et al.*, 2010).

A place of situated learning

Yearning for the countryside creates an impetus for revisiting and recreating heroic acts, while clarifying how social imaginaries interact with media portraits and nostalgic discourses cast the landscape of rural Guizhou as the nation's past. The young travellers' narratives frequently reflect the legacy of the 'zhiqing' generation – learning from the rural situation and helping society. The wish to learn indicates that tourism may go beyond the motive of personal discovery (Li, 2000), reflecting a disturbing sense of spatial alienation from the legacy of the rural that is perceived to be the cradle of Chinese civilisation.

Tourism thus exhibits tourists' recognition of individual and national rural heritage. For the young people who have spent their whole lives in cities, travelling to rural Guizhou affords an opportunity to materialise connections with a meaningful source of their cultural and spiritual identities. The countryside is therefore identified as purer, acting as a cleanser for the urban tourists who, through the spatial movement of travel, can see through the mist of materialism that veils the urban existence, to build an existential connection with their spiritual selves, as a young backpacker stated:

> I've always dreamed of being a volunteer. I like the simple life. I don't need much stuff anyway. Life in the countryside is less stressful. People are friendly here, and they live with the best scenery. I like living here because it's more basic and less things to worry about.
>
> (Direct interview quotation from field research)

This young backpacker's perception of the rural human and natural environment reflects a sense of alienation from and frustration with the prevalent modes of urban life in which capital pursuit has created environments that are detrimental to individual wellness (Vidon, 2019). Like the 'zhiqing' group of older tourists, the young travellers regard cities as places where materialism prevails. Only in the countryside can they find a space for cultivating the 'true' self. The young people's narratives reinstate the idea of the rural as a place that is somehow more authentic than the inauthentic urban; rural life is harder but purer, human relationships are more meaningful and honest:

> We come to the countryside to get close to nature and see free animals, such as some horses or some cattle. We want to see some different people and things... just interest really, see how they live, we want to see their customs, eat fresh organic food, learn more about nature. (Direct interview quotation from field research)

Young travellers' association of the rural and the natural is manifested as an irresistible temptation of difference between rural China and its urban counterpart. It marks a facet of critique against modern environmental degradation. Since the 1990s, global environmental discourse has entered Chinese society. The environmentally aware citizen is thus characterised as a green consumer, deeply concerned with the ever-worsening environmental degradation on individual lifestyle and health choices (Farquhar & Zhang, 2005). Both the older tourists and the young travellers believe that the rural environment is cleaner and purer than cities. Appreciation of environmental problems and their association with individual health are translated into the practice of the young travellers. When travelling with a group of university students on a geography field trip to rural Guizhou, the researcher found that they envisaged rural Guizhou as spatially beyond, and temporally before contemporary problems of environmental degradation and pollution which are seen to characterise urban China:

> We like to come to this kind of place because our cities are very polluted. The air is clean here, all food is fresh. If the peasants want to eat this vegetable, they take that vegetable and cook it. If they want to eat pork, they kill the pig and eat it. Their life is much healthier than ours in cities... When we eat, we don't know what we are eating, there are so many fake

products in China, and many of the foods contain unhealthy chemicals. But in the countryside, the food is simple and healthy.

(Direct interview quotation from field research)

The students' narrative indicates the way in which urban tourists associate spatial movement from cities to the countryside with individual health benefits and their own environmental awareness. This dovetails with the idea, now being popularised by the national slogan: clear clean rivers and lush green mountains are invaluable assets. The countryside is thus regarded as a place of harmony and health, translated into the choices of tourists and their perceptions of rural China. Not only does travel to the countryside represent an opportunity for tourists to enter 'cleaner' environments, but it also acts as a mechanism whereby consumer choices are used to portray the individual as eco-friendly (Farquhar & Zhang, 2005).

For the young travellers, the countryside offers an environmentally utopic space to escape from the contamination of nature caused by the unbridled pursuit of capital gain that is prevalent in contemporary Chinese cities. Travelling to the countryside has become a form of critique through the medium of spatial transition whereby urban people are able to embody their dissatisfaction with what they see as the ills of an increasingly urbanised China. By so doing, they seem to feed directly from the cultural reservoir of socialist nostalgia that is institutionalised by the Chinese nation-state through genres of public education including tourism (Li *et al.*, 2010). Mediated through rural tourism, both young and older generations attempt to reconstruct the past, in an idealised fashion, to act as a stabiliser in an uncertain present of post-reform China.

Conclusion

The countryside is imaged as a place of tradition, bounding rural communities against the uncertain nature of the cities that are continually changing under the force of China's modernisation campaign. The urban tourists are driven to visit rural Guizhou through impulses aroused by the differences defined in the national discourses of the rural–urban separation. These same differences also feature the practice of rural tourism, wherein the cultural values associated with the countryside and cities are encoded in the behaviours and perceptions of urban tourists, in turn channelling identity politics between both urban and rural places and peoples.

One would argue that the processes of social-cultural change direct the urban gaze towards the rural, idealise it and consume it in a manner that reinforces expectations, despite that the objective realities are far from ideal. Although rural Guizhou is imagined as a peripheral place, the very fact that tourists target it is thanks to its incorporation into wider national developmental networks that allow urban tourists to venture further to find more romanticised versions of rural China. Ironically, tourist discovery of rural Guizhou is not incidental, and occurred as a result of government planning that drew the place into wider networks of China's modernisation campaign. Those processes will become subject to more development intervention, changing the previous, the current and the future characteristics of the place.

It is perhaps inevitable that tourism will penetrate rural Guizhou, when an increasingly globalised China emerges. Campaigns are under way, including the 'Western Development' (Chang & Wang, 2020) and 'Poverty Eradication 2020' (Cao & Hu, 2020). They are processes that will continue, in material and communicative terms, and in terms of how rural Guizhou and the people are presented, imagined and consumed (Tian, 2020). The tendency of rural Guizhou to absorb increased tourism will become a seminal nexus of how the local communities come to identify themselves in contemporary China. Urban tourists visit rural Guizhou precisely because of its imagined fixity against change, whereas conversely the local people hope that the expansion of tourism brings developmental change and pushes their 'bounded' place into wider communicative and economic networks.

Through the perspectives of Chinese domestic tourists, this chapter has delineated the ways in which they consume the perceived traditionalism and fixity of rural Guizhou. Despite China's institutionalised narratives of rural and urban differences, this chapter shows that tourism has become a space of reflexivity. Davis (1979) argues that collective nostalgia draws on contemporary social-cultural and institutional conditions to formulate expectations, norms, behaviours and values. Older tourists of the after 50s'–60s' generation are probably best placed to contemplate China's changing social-cultural and economic landscapes. They lived through a dramatic U-turn in the dominant ideology of China, shifting values from revolutionary-socialist-idealist asceticism to be replaced with the current dominance of global capitalism and the growth of materialism. Those changes have very real individual consequences: social security has been eroded while uncertainties have emerged under market influences. For that 'zhiqing' generation of urbanites who were fed from the former secured employment, the current reform era of social-cultural change brings some disturbing personal concerns.

Young travellers perceive the countryside through 'utopian nostalgia' – coming to rural Guizhou to help the less fortunate in society. Their actions invoke the utopian dream of the pre-reform era – obedient people and a caring and righteous government working together to build a socialist paradise. It is wishful thinking to make an 'incorruptible' society and a 'caring' government operate in unity to enable China to rise (Rofel, 1999). Mediated through tourism, both the older tourists and the young travellers drew a direct lineage to the narratives of the past, reaffirming the discourses that overcoming hardships would help formulate and nurture preferred moral characteristics. Nostalgia thus becomes symptomatic of wider social dialogues through rural tourism, suggesting the ways in which people feel about the present. However, nostalgia is not the privilege of social elites, and it may be regarded as a resource that social groups exploit for their own sake. This chapter demonstrates that nostalgia may be tactically employed as a mechanism to critique contemporary society, transforming touristic nostalgia into a keyway for disparate peoples to connect in China's new geographies of movement.

China's modernisation campaign directs urban people's attention to idealised rural places, while the 'primitive and backward' lifestyle of the countryside inspires expectations and reflections on the objective reality. China's existing dichotomy of urban and rural discourse defines modernity as a spatial geographical centre versus periphery concept in the relativity of 'progress' and 'backwardness'. However, cultural significance and power relations are the keys to creating a space of cultural geographies. The related analyses should move beyond both the environmental perception studies that focus on individuals and realism's geometry, towards the regularities of intersubjective images of place that are imbued with emotional content and historical significance (Shields, 1991). In this vein, this chapter verifies that understanding human nature is essential. Tourism enquiry should not only place an emphasis on people's experiences, but also place a more abstract focus upon their ability to understand the world – the human consciousness. The existing analyses of tourism within China tend to concentrate on the mechanism and content of state policy and management. In contrast, the interrogation of tourist experience presented in this chapter has achieved an alternative goal: exploring the possibility of pushing tourism enquiries of China from a positivist spatial science seeking objective facts, to a more humanistic field of research.

Conclusion: Applying Ethnography to China Tourism Research

Introduction

In China, tourism is often introduced to the public as an economic development mechanism, but its role is not only in the economic aspect. Tourism is multifaceted by nature as both an industry and a practice (Mostafanezhad & Swain, 2019). Tourism enquiries manifest three methodological paradigms: the scientific positivist, the interpretive and the critical (Tribe, 2001). The scientific-positivist paradigm largely dominates tourism research. Especially within China, this paradigm ensures that the analyses of tourism concentrate on the mechanism and content of the nation-state policy and management, guide development projects with 'positive' information about the modern Chinese nation, which transmits official narratives involving history, cultural heritage and economic development (Bao *et al.*, 2014; Li & Hu, 2008; Li *et al.*, 2010). This tendency leaves a critical issue to be resolved, namely multi-fronted research alternatives are needed to progress and diversify the tourism research landscapes inside China (Hollinshead & Hou, 2012). However, according to the author's review of the latest documents (Academy of China Social Science, 2022; National Natural Science Foundation of China, 2022), this issue remains unresolved. So, one of the purposes of this book is to unsettle the deterministic approaches that dominate the academic sphere of tourism research related to China, while responding to the call of scholars for creating spaces where diverse voices can be heard (Ivanova *et al.*, 2020).

Adopting a humanistic approach and relying largely on the notion that place is the centre of meanings, this book has so far attempted to offer a timely intervention to understand the dynamic role of ethnography in tourism-related enquiries of people, place and society. There is no consensual definition or single procedure to follow when using ethnography, attributable largely to the re-contextualisation of ethnography over time through diversification in and across many disciplines (Lambert *et al.*, 2011). Therefore, it is imperative that tourism researchers consider some trademark features of ethnography in the related tourism studies. Accordingly, a theoretical framework is developed especially for the

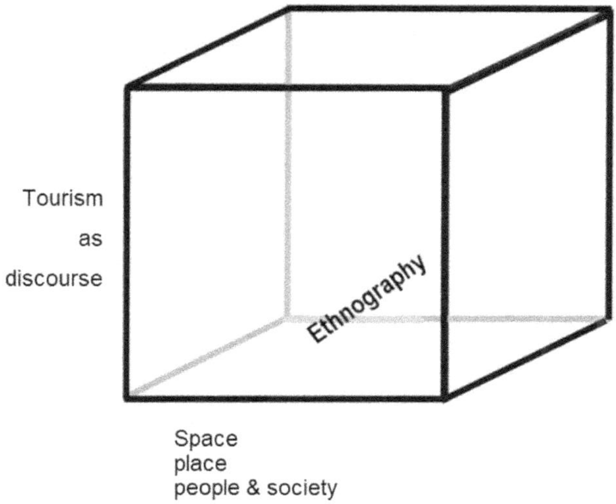

Figure C.1 A three-dimensional research framework (Source: Author)

enquiry of tourism and social-cultural change in China. This framework regards tourism as a discourse of difference, and from this point of departure to explore the tourism phenomenon based on three perspectives: the discourse perspective, the spatial perspective and the ethnography focus (Figure C.1).

The discourse perspective

Tourism may be regarded as a discourse of difference through which power is propagated by the narratives of practice. Power is exerted through systems of thoughts composed of ideas, attitudes, courses of actions, beliefs and practices that construct the subjects and the worlds of which they speak systematically – in other words, through the construction and propagation of normalising discourse (Foucault, 1980). Discourses constrain individual agency by shaping social relations and identities and the overall contours of the taken-for-granted world, which both naturalise and universalise people's worldviews (Thorpe, 2012). Discourses emerge from systems of knowledge that help to shape explicit understandings of the world through cultures along with customs, organisations and individuals who create networks of influence and specify power relationships.

Foucault (1980: 51) argues that power is enacted by the dissemination of certain kinds of knowledge to formulate our ways of seeing the world: 'the exercise of power itself creates a condition for the emergence of new objects of knowledge and accumulates new bodies of information'. The creation of discourses enacts and cements power implicit in all kinds of relationships (Foucault, 1990). Tourism may be considered one such medium infused with certain ideologies, images and messages that signify social-cultural, economic and regional differences about people and places (Hollinshead & Hou, 2012; Lau & Li, 2019; Li & Lau, 2022). Tourism representations of reality passively reflect the prevailing social-cultural values by drawing upon existing stereotypes and images (Hollinshead *et al.*, 2009; Overton & Murray, 2016; Pan *et al.*, 2017; Zhang *et al.*, 2015). It thus plays an important role in power distribution, shaping values, behaviours and identities due to its contribution to the process of socialisation (Lau & Li, 2015; Morgan & Pritchard, 1998).

Tourism may be used to represent certain versions of the world at the expense of others. In tourism, some voices are heard more than others, certain histories are presented more widely than their counterparts and a few narratives are selectively used to recreate histories of people and places. For instance, contrasting images of communism are selectively presented in the practices of communist heritage tourism by Eastern Europe and China (Caraba, 2011). Tourism selection of discourse types is largely tied with institutionalised worldviews and ideologies (Ashworth, 1994; Zhang *et al.*, 2015). The practice processes of tourism, therefore, have broad social-cultural meanings that extend far beyond the tourist consumption of places, while providing an instrument by which a nation-state can project ideological messages to legitimise its position (Li & Hu, 2008; Li *et al.*, 2010; Li & Mai, 2022; Zhang *et al.*, 2015). Especially in tourism target areas where disparities in power exist between tourists and toured communities (Bruner, 2005; Caton & Santos, 2007), dominant groups steer developmental policies to target less powerful peoples, forging influences in tourism representations of peoples and places (Oakes, 1995, 1998, 2016; Tresidder, 1999).

The spatial perspective

The spatial dimension deals with the specific assets, constructions and perceptions of a place (Lau & Li, 2015, 2019; Li & Lau, 2022; Scannell & Gifford, 2017). The concept of place captures the meaning of a location for the individual, referring to a 'distinctive coming together in space' (Agnew, 2011: 317). Perceptions of and interactions with places reflect the ways people understand the world in which they live, and help them to determine what is, or is not, important (Cresswell, 2007). Being profound centres of human existence, places are keys to connecting all human feelings to the material environment (Relph, 1976; Tuan, 1975, 2001a, 2001b). Places are significant because 'they are the focus of personal feelings… [and those] feelings given to places may be so strong that they become a central part of the identity of people experiencing them' (Rose, 1995: 88). However, places can be changed by forces beyond human control, thereby raising a series of disturbing questions as to what happens when places change.

Harvey (1996) investigates those questions, pointing out that the advance of communicative technologies has, in many ways, made people feel that their places are under the threat of change. He uses the term 'time-space compression' (Harvey, 1989) to describe the key features of the postmodern society where the rigidities of the economic systems are succeeded by the ever-expanding forces of global capitalism. Similarly, Giddens (1990) uses the concept of 'time-space distanciation' to clarify how facets of social reality and interaction are extended across time and space. In this vein, individuals' social-cultural realities are constantly shaped and reshaped beyond their immediate physical location in space and time. Both scholars predict that the advance in communicative technologies is to connect disparate places across geographical distance, increasingly the interlink between human societies.

Massey and Jess (1995) put forward a so-called 'global sense of place', as a new theoretical framework for characterising places as processes instead of sites of isolation and localism. For instance, the emerging global cities along China's coastal provinces manifest convincing explanations of the processes that this new framework is conducive to change; and demonstrates that outside influences are initiating waves of locality reorganisation, challenging the theoretical hegemony of isolated places. However, this framework does little to address human feelings of uncertainty that are brought about by globalisation, nor does it consider the shifting attitudes of people towards changes in places. Consequently, an issue remains: this framework does not help tackle the impending threat to both personal and group place identities enforced by constant alterations to the physical, social-cultural and emotional landscapes of places that people call home.

The ethnography foci

Tourism is a modern and postmodern phenomenon, and is considered one among the changing forces of contemporary human society (Higham & Miller, 2018). It is essential that tourism researchers focus academic attention on human nature to study 'man-in-the-world'. Scholars maintain that tourists are place attached, and this attachment influences their satisfaction with and interest in a destination place (Abou-Shouk *et al.*, 2017). Ethnography is defined as the systematic study of people and cultures (Hammersley & Atkinson, 2019). It offers an effective approach to the study of tourism. Essentially, ethnography is a field-orientated activity that has cultural interpretations at its core (Hunter, 2020). It is a humanistic approach designed to explore cultural phenomena where the researcher observes society from the point of view of other human beings – from the point of view of the subject of the study.

A human being has a fully centred self, able to transcend any given environment in the direction of her/his world (Tuan, 1965). Humanistic enquiry sees three standpoints existing as human positionality: as a member of the biological species; as a member of a sociocultural group; and as an individual (Tuan, 1967). From each standpoint, different dimensions of a human worldview may be considered, including the human body (the species), gender and cultures (the sociocultural group) as well as personal experiences and history (the individual). The humanistic enquiry of tourism is not only to place an emphasis on people's experiences, but also to place a more abstract focus on their ability to understand the world – the human consciousness. More than being a mere location, place has both symbolic and experiential meanings that can be uncovered by exploring the perspectives of the people who 'create' them through personal experiences such as tourist activities.

Humanists value people's experiences and self-consciousness to understand how they perceive the world. Accordingly, the cases and projects that are presented in the chapters of this book attempt to explore tourist understanding of the places in China as the lived environment by relying largely on the notion of place as the centre of meanings, through the lens of tourists' experiences and host–guest interactions in places. For instance, the physical and social-cultural distances of China from the rest of the world functioned as a key attraction to entice the international tourists during the last two decades of the 20th century (see Chapter 1). Due to its remote location and density of ethnic minority populations, Guizhou Province is imagined as a true example of the key social-cultural and economic distance that typifies China's rural–urban separation. Chapter 10 delineates the ways in which the countryside becomes transcended through the mediation of tourism, rendering sublime the globalised city lives of the Chinese domestic tourists. In so doing, the chapter demonstrates the effective use of ethnographical methods to uncover how travel experience connects rural and urban peoples through the processes of tourist imagination and consumption, and how the contemporary discourse of tradition fuses with individual histories to formulate the ways in which Chinese urban folks understand rural China.

Broadening Ethnography

The three-dimensional framework provides a broadening perspective for ethnography. It posits tourism as a discourse of difference, offering an ideological prism through which to examine the realms of the sociocultural and spatial politics in China's contemporary growth and change. Thus, the extent to which those discourses are accepted and subverted by both host and guest communities, and the ways in which they shape material and lived realities are contemplated in reference to Structuration Theory (Giddens, 1984) when ethnography is operationalised (also see

Figure 9.2 in Chapter 9). This theory guides contemplation of how tourism is enacted from the nation-state level, received at the society level, and then how social action feeds back into the material properties of places in China, as is manifested in red tourism development (also see Chapter 5). Under the formulation of the tourist gaze (Urry & Larsen, 2011), the power of discourse in tourism is enacted at the social level, creating a dialectic between structure (state-planned tourism) and agency (its receipt at the societal level). The rationale is to interrogate more rigid dualisms between structure and agency by examining the state mechanisms of place and subjectivity creation on the one hand, and individual understanding of the self, society and place on the other. In so doing, the chapters of this book probe the extent to which tourism is a medium through which individual tourists and hosts both create and recast their own subjectivities.

Three interrelated considerations

Agents often face intractable structures in the form of rules, social classifications, gender and race issues in their day-to-day lives (O'Reilly, 2012. The perception of a duality between structure and agency is heightened by the fact that human action is short term whereas social structures appear more embedded across space and time. To underpin the overall research strategy, it is proposed that this idea be adopted – the recursive relationship between the duality of structure and agency in which each component is researched through empirical evidence. In different social situations, agents are placed in dynamic relationships of position practices (practices that occur in relative positions), placing a focus on agents who are situated in webs of networked relations which need to be considered in terms of structures and agents in particular contexts for empirical research (Greenhalgh & Stones, 2010).

In Chapter 9, Figure 9.2 suggests that empirical research should attempt first to sketch the structural context of the field to ascertain the most salient explanatory forces. Upon this, a framework of the intermingling influence of relative domains should be outlined along with considerations of which actors-in-focus can be conceptualised to create a wider field of study. Methodologically, a process is created whereby the researcher works backwards from the phenomenon in question, attempting to find the causes of its production. The cases in this book situate placed agents, including hosts and guests, in the contextual field of place, ethnicity, landscape, people and tourism. The intermingling fields in which agents are situated are analysed through scrutinising textual tourism promotional materials and tourism-related policy, as well as the spaces of tourism that are considered material texts through which power relations can be discerned. Foucault's concept of the omnipresence of power sets the understanding of the flows and exertions of both social and institutional power as it is enacted, absorbed, confronted and transformed in the day-to-day practices of life in China.

In sum, the central research questions that are manifested throughout the chapters involve the creation of tourism in China: how are the places created, and why? In the context of tourism, differences between agents' communities are not only institutionalised, but are also emphasised on a daily interactional level. In this vein, three interrelated research considerations are highlighted through the chapters in this book: (1) place – an account of the social production of space; (2) society – an understanding of the ways in which agents interact to produce and reproduce social relations and places; (3) power – an outline of the means through which power is made inert in and enacted through social relations, forms of knowledge and places. Those considerations enable a holistic contemplation of the picture of how tourism is enacted from China's nation-state level, received at the society level, and then how social action feeds back into the material properties of place. Eventually, three methodological techniques are formulated. First, tourism policy and promotional materials are considered with discourse analysis. Second, the bulk of empirical investigations are conducted through fieldwork that utilises interpretive techniques centred on interviews. Third, non-representational techniques such as observation of interactive participants are used as supplementary data collection techniques.

Operationalising Methods of Ethnography

Ethnography is utilised mostly to investigate the cases analysed through the chapters of this book. Evidently, it is an effective methodological approach to enable comprehensive observation of tourists' experiences and perceptions of life in China and recording of those experiences and perceptions. Above all, it is believed that ethnographical interpretation and observation not only helps to reflect, formulate and respect the creativity of social practice, but also provides researchers with critical explanation for such practice (Burawoy, 2003).

Social and cultural interpretation

According to Morishita (2001), a researcher should become deeply immersed at the site of investigation to observe and interpret the phenomena in question, because social activities can be read for meanings just as spoken and written words are commonly read. Deep immersion ethnography thus regards the researcher as the vessel for interpreting an assemblage of texts, which enables them 'to be contextually sensitive, inductive and naturalistic' (Patton, 1982: 10) towards the subject of investigation. This type of approach expects the researcher to become a key influence on the creation of knowledge through interpreting the words and actions of others (Jamal & Hollinshead, 2001). However, concerns may arise that some social and cultural interpretations that researchers conduct are beyond the realm of cognitive consideration for most lay-people. To resolve those concerns, non-representational approaches are recommended, such as participant observation.

Social and cultural interpretation may be understood, arguably, as a meaningful social activity. The key is to be sensitive to the context to summarise it (Morishita, 2001). Human consciousness shapes social and cultural practice in both discursive and non-discursive ways. Ethnographical fieldwork, therefore, requires that a researcher mine the field to capture and understand non-discursive information emerging in the investigation process (Latham, 2003). In general, social and cultural practice manifests three aspects of human consciousness (Giddens, 1984): (a) discourse awareness that explains what individuals do; (b) practice consciousness that determines beliefs and knowledge of oneself and others; and (c) unconscious thoughts that drive the 'natural' behaviour of 'instinct' or 'common sense' (Figure C.2). For the discursive and practical aspects of consciousness, methods such as interviewing enable respondents to discuss their worldviews and feelings. On the other hand, observation helps researchers to understand that culture may be about what people do in their daily lives.

Appreciation of those aspects marks the researcher's recognition that social practice is embedded in ways of being in the world, which are rather different from discursive ones – this notion correlates with Giddens' (1984) unconscious motives/cognition. Those embodied routines are not ordered through the discursive, thus they require the use of methodological tools that are also not discursive (Larson, 2009). Accordingly, during the interviewing process, a researcher is supposed to let the respondents naturally relate their personal feelings, in order to interpret both conscious discourse and practice, while observing the feelings that the respondents may not be able to express for various reasons (Jacobson

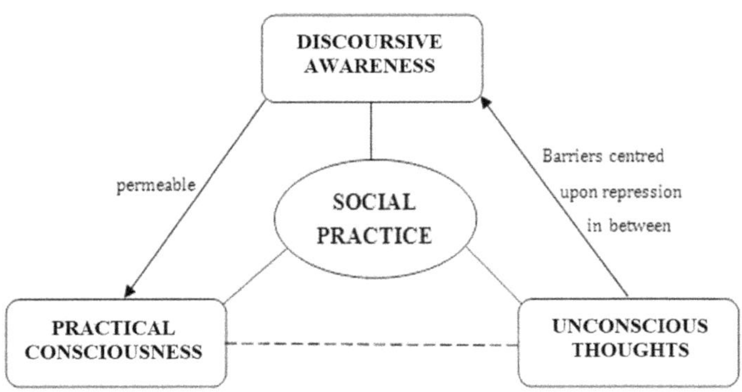

Figure C.2 Manifestation of human consciousness in social practice (Adapted from Giddens, 1984)

& Larsen, 2014; Larsen, 2014). There are three reasons why methodologies that consider the flow of practice and its complex embodied intersubjectivities can be useful.

First, some interview questions about respondents' choices may be far beyond their ability to ponder in depth during their everyday lives. Second, that kind of question may demand cognition without any association with the ways the respondents usually go about making the choices they do in their daily lives. Instead, a researcher can 'gain a sense of the frame of reference through which individuals encounter and negotiate their world by accounting for practice' (Latham, 2003: 72). Third, the reason for respondents' inability to articulate some practices is that they come from the knowledge accumulated by straightforward actions that make certain behaviours intuitive. To access reasoning behind them, a researcher must go beyond the discursive. By observing the practice of everyday life, ethnographical interpretation can embody, enact and respect the creativity of the practice while still offering useful (and critical) accounts of it.

In a nutshell, the fieldwork of those research projects presented through the chapters of this book is designed through adapting to possible everyday life situations. This strategy not only ensures that both conscious and unconscious behaviours and discourses emerging in the process of interpersonal communications are meticulously observed, but also enables a holistic interpretation of culture and society. For instance, the focused fieldwork of the study presented in Chapter 10 concentrated over at least three months, but the initial contacts with the investigation sites had begun as far back as several years before. The people and places of China change every year under the nation-state planned modernisation campaign projects (Chang & Wang, 2020) and 'Poverty Eradication 2020' (Cao & Hu, 2020). Thus, further brief visits were made to the investigation sites to observe the most recent changes in the studied cases. Over 15 years, the author strived to witness as much change as possible, and attempted to establish trust and relationships with the local communities of that specific area in one of China's poorest provinces – Guizhou.

Access to respondents

In most of the cases analysed in the book, the author presents how qualitative methods are utilised to gain access to respondents who are either international or Chinese domestic tourists, local community residents, government officials or tourism industry practitioners. In Chapter 10, for instance, two specific types of populations are targeted for investigation: older nostalgic tourists (after 50s'–60s' generation) and young idealised travellers (after 80s'–90s' generation). The former is one of the best-placed generations to contemplate China's social-cultural, economic and physical landscape changes that have taken place since

1978 when economic reform started. The latter, as direct descendants of the former, are best placed to feed from the life experiences of their parents' generation, to review the fast-changing urban landscapes and ponder the transformation of China's current rural–urban relationship. In this case, determining relevant respondents was dependent on an internet search. Updates were carefully watched of several official websites among agencies that launch, several times a year, commercial package tours or organised volunteers travels to rural China, including travel companies, photographer associations and volunteer organisations.

As the fieldwork concentrated on frequent and focused research trips over many years, it allowed the researcher time away from fieldwork activities to develop the research ideas from a large pool of field data and relevant readings. A critical challenge of ethnographical fieldwork is that researchers should be able to remove the personal self from the immersed nature of the field to intellectualise what is seen and heard, and put it into perspective (Islam, 2020). So, researchers should physically leave the fieldwork areas for extended periods of time, to return afterwards with clearly developed ideas of research topics and questions. This strategy would help identify and diversify, at different times, the target respondents.

In most of the cases presented in this book, the researcher would make shorter visits to the sites of investigation to gain access to the respondents, and gradually deepen and widen the investigation. Firstly, shorter visits enabled the researcher to have a more detached perspective, be embedded and friends with potential informants. Shorter visits provided an equilibrium of knowledge about both tourists and the local communities. Moreover, periods of 5–10 days were considered conducive to the general practice of the tourism business, offering sufficient time to observe the flows of tourists under investigation. It is important that the researcher never overstayed the welcome period with the local communities which, due to the peak tourism season when the fieldwork visits were made, often appeared very busy. As the fieldwork progressed, the researcher became more interested in the tourists' perspectives, and increasingly interactive with both the tourists and the local people. Gradually, the researcher found more visitors in 'holiday mode', characterised by high spirits and openness to meet new people, and to cooperate with the researcher.

Interview and observation

During fieldwork, researchers should be aware of these issues: inappropriate access to participants, over-representation of the self-experiences and questioning fundamental parts of self-identities (Islam, 2020). In the situation of China especially, this awareness requires that a researcher look beyond the Western-centric conceptions of ethics during field research, because concepts such as confidentiality and benefit are appreciated very differently across cultures (Kindon, 2002; Pain, 2004; Randstrom & Deur, 1999). Local people may be suspicious of researchers from outside China, whether they are simply 'dirt-digging' – doing fanciful research and not able to answer the important questions of the Chinese mainstream (often more positivist) research (Li & Mai, 2022). In such circumstance, researchers may have to use recorded interviews and jot down detailed information, or simply use short voice notes to record certain information that could be important, especially in informal or random interactions with respondents, in order to avoid any unforeseen consequences. In most cases analysed in this book, two techniques were predominantly operationalised in the field investigations – face-to-face, unstructured interviews and participant observations.

Unstructured interviews allow the respondents the opportunity to put together their answers naturally, since interviewing itself is to understand how individuals make sense of their own lives (Sullivan & Forrester, 2019). In the case analysis in Chapter 10, for example, unstructured and informal interviews were conducted on both older tourists and young travellers, with topics including the impacts of tourism, tourists' perspectives and some wider issues regarding local communities and Chinese cultures. More traditional 'deep immersion ethnography' and short-term 'rapid participatory assessment' methods were also utilised (Chambers, 1991). Both techniques enabled the researcher to understand the interaction mode between the local community and incoming visitors, as well as to observe how tourism impacts the place of the target research sites.

In the case of Chapters 1 and 10, recorded face-to-face interviews accounted for only a small portion of the fieldwork data, because social practice is rarely ordered through the discursive (Bernard, 2006). Many more conversations became footnotes to the fieldwork, helping snowball research ideas, test and verify theories, as well as find potential respondents for more formalised interviews. Therefore, it was essential that the researcher look for repeated themes in informal conversations, trying to sort out topics that people brought up. Those topics inspired the researcher to conceive questions aligned to the dominant discourses, narratives and impressions of the respondents. Observation thus helps to understand patterns of interactions among tourists and local residents, their cultures and where and when tourism infiltrates the lives of the research subjects. Observation is also used to construct information on the physical aspects of the local landscapes associated with tourism development (see also Chapter 9).

At times, fieldwork researchers may find themselves complete participants, participant observers and complete observers. So, they must keep their personal selves aware of the different approaches they are to employ. 'Touristic border zone is regarded as a local place wherein the goings-on are determined by global forces that congregate there' (Bruner, 2005: 17). So the quintessential 'outsider' exists even less in tourist places – those of flux and mobility – where unfamiliar faces are taken for granted. In this vein, it is relatively easy for the researcher to join tourist groups as a complete participant or predominantly as a participant observer. Evident through the chapters in this book, the researcher shared 'outsider' traits with other tourists (see also Chapter 1), and indeed, coming to China's geographically remote rural Guizhou, for instance (see also Chapter 10), the researcher as a city person would frequently be seen by tourists and local people as an outsider similar with other tourists. The researcher must frequently make the effort to disassociate themselves whenever necessary.

Evident in the case analyses throughout the chapters in this book, the researcher mined the field, and recorded the observed information in field notes, using them to mark down doubts, fears, concerns and feelings towards the phenomena in question (Cook & Crang, 1995). The researcher found the fieldwork exercise surprisingly immersive, and thus became deeply engrossed, from time to time, in certain situations that, reflecting after each specific event, would allow a better understanding of the conversations or observations immediately afterwards. Daily self-reflection enabled an evaluating mechanism that helped track any possible arrival of converse information, thereby questioning the previous suppositions, to reaffirm the need for personal detachment. Daily recordings of personal feelings towards individuals, situations, methods and happenings functioned in retrospect which helped assess the fieldwork data. The recorded on-site observation invariably implicated selection of information. Therefore, such observation should not be regarded as a process of passively recording 'fact'. Instead, the selection of field information that should be recorded and how it should be recorded manifest a process of interpretation and evaluation of what was or was not significant. The recording of on-site observation is a key step to turn events into accounts that present objects, people and situations in certain ways (Emerson *et al.*, 1995).

Physical mobility proved necessary during the fieldwork, for visiting different places across China. Multimedia methods thus provided a means that assured instant recording of thoughts and information, as well as the integration of the actual phenomenon, subject or scene that was being recorded, interpreted and described. Retrospection on scenes and images aided reflecting and understanding what was observed, due to the intensity of the fieldwork, and the gap between the sociocultural and physical characteristics of the investigation sites, as well as the researcher's own world. It has been found that multimedia methods are effective for recording observations in a touristic situation where the local people are more accustomed to having cameras pointed at them and their surroundings. This kind of situation would usually enable a researcher to make video and audio notes with ease.

Some field research highlights

The flow of field research was three-fold, manifested especially in the case analyses in Chapters 1–5 and 8–10. Firstly, the respondents were invited to select interview topics, and they would usually choose to discuss some daily life topics. Secondly, the respondents were invited to answer questions unrelated to daily life that needed recognition, because their interpretation of a personal experience might function as a frame of reference to help analyse the respondents' inner world (Jamal & Hollinshead, 2001). Finally, when a respondent had difficulty clearly explaining the reasons behind some behaviours, the observation technique was used to visualise those behaviours, helping infer what kind of experiences those behaviours might have accumulated. All in all, the methods of ethnography have proven effective in constructing the following information for exploring the tourism phenomenon in China:

- Tourist experience and host–guest interaction: The lives and activities of tourists, entertainment personnel, handicraft vendors and community residents, as well as interactions between different groups. In addition, it is essential that both the surrounding non-tourism areas and their life and mode be closely watched and deeply understood, to measure the impacts of tourism on the destination proper as well as its surrounding communities.
- Physical landscape and features: Tourism infrastructure construction and its location, introduction and form, natural landmarks, different architectural styles and specific tourist attractions marked as 'modern', 'typical', 'traditional' or 'authentic'. Related to the previously listed items, it is also important that styling, especially the dominant style, is carefully considered, as it is regarded as the basic version of 'signs' (Urry & Larsen, 2011).
- Spatial layout and characteristics: The economic, social-cultural and material characteristics of scenic spots. Those enable people to distinguish between the perceived space and the physically segregated spaces in the tourism zone (Edensor, 2009) and to understand how the surrounding environment of the tourism destination can be successfully protected, to prevent potential damage from excessive development.
- Impact of resource commodification: Examination of the ways in which landscapes and community cultures are packaged for consumption, including an analysis of discourse surrounding the actual presentation and consumption of resources.

- Space of tourism: The ever-changing spatial dynamics of tourism destination places, in association with the development and management of new infrastructure, tourism segregated spaces and architectural features.
- Globalisation and connectivity: The impact of tourism and its internal and external linkages associated with the local economy, such as changes to, and the consequences of economic features enforced by tourism; interaction between internal and external groups in a tourist destination place. Additionally, social structure modifications should be observed because they relate to the global impact of tourism in times of digitalisation, on the role of the traditional labour force and capital, as well as the change in the population structure of a given society.
- Government intervention: The respective roles of government and non-local businessmen, as well as hosts and tourists, in making development policies and strategies to determine how the tourism business should be managed and what cultural aspects are appropriate to present to tourists.

Conclusion

The above list for exploring the tourism phenomenon in China should not be regarded as exhaustive because multiple meanings of place are often centred on tourists' perception of and interaction with specific places, as uncovered through the chapters in this book. Current tourism development affixes both past and existing economies, as well as cultures to places of China, like a force of change that interacts and integrates with parts of new realities. To consider the varied imaginative, material and performative territory at the heart of the findings that are presented, this chapter presents a three-dimensional framework that accounts for the interaction of power relations, a humanistic perspective of place and individual practice. Its application enables the interrogation of the dialectic distance between places in China. Against the ever-changing social-cultural landscape of the country, the rural and remote are frequently imaged as places of tradition, fixing the communities there against the uncertain nature of China's cities that are changing under the forces of the nation's modernisation campaign. International tourists were enticed to visit the 'distant' China during the early days of the nation's opening-up at the turn of the century; so were China's urban tourists driven to visit the countryside, through impulses stimulated by the differences which are defined in the global and national discourses of differences. Those same differences also characterise the practice of tourism in China, wherein the social-cultural values associated with nation, society, place, countryside and city are encoded in the behaviours and perceptions of tourists, in turn shaping identity politics between both the places and peoples of China.

The book presents specific case analyses to demonstrate that the processes of social-cultural change direct the urban gaze towards the rural, idealise it and consume it in a manner that further mobilises it to reinforce expectations, despite the fact that objective realities are far from being ideal. For instance, rural places are imagined as peripheries of the nation, and the very fact that tourists target their selected rural places is thanks to their integration into wider developmental networks that facilitate urban tourists to venture further in order to find more authentic versions of China that are still, relatively, undiscovered by tourists. The irony is that tourist discovery is not incidental, and occurs as a result of nation-state planning that draws rural places into wider networks associated with the mechanism of China's modernisation campaign. Those processes will become subject to more development intervention, changing the previous, the current and the future characteristics of China.

As this book neared fruition, international tourism began to see a ray of hope: more and more countries are reopening their borders to welcome tourists after more than two years of a global pandemic crisis. However, tourism within China started to face even more harsh setbacks. Since January 2022, the Chinese central state has tightened its so-called zero-COVID policy, making recovery of the tourism business slow and hard. Figure C.3 (from left to right) shows (1) a QR code at the entrance to Zhuhai City Resort Island Ferry Terminal for passenger health check; (2) the wharf surrounded by an iron fence that is used to keep away potentially unchecked passengers; and (3) a tourist space appeared empty during the seven-day National Day Holiday of 2022. Lockdowns and unpredictable restrictions across China have led to soaring mental health problems, and the human cost can be measured in stress and anxiety, rather than holiday relaxation during this so-called golden-week peak tourist season. Many, including the World Health Organisation (CNBC, 2022; Millard, 2022), criticise China's strict COVID control measures as 'inhumane' and 'unsustainable'. In this circumstance, holding a

Figure C.3 Landscape of the new normal under the zero-COVID policy (Source: Author – photos taken 2–6 October 2022)

humanistic perspective to examine China, particularly tourism development, appears unprecedently necessary.

Holding a humanistic perspective and relying largely on the notion of place as centre of meaning, the author's application of ethnographical methods aims to unsettle the deterministic approaches that still dominate the academic sphere of tourism research related to China. The case analyses through the chapters in this book are to encourage the reader to realise that the existing dichotomy characterising China's nationalistic tourism discourse defines modernity merely as a spatial geographical centre versus periphery concept in the relativity of 'progress' and 'backwardness'. On the contrary, the author would argue that cultural significance and power relations are the keys to creating the space of cultural geographies. Associated analyses of people, place and tourism in China should look beyond both environmental perception studies that focus on individuals and realism's geometry, towards the regularities of intersubjective images of place that are imbued with emotional content and historical significance (Shields, 1991). In this vein, this book reaffirms that understanding human nature is essential and, to achieve this, ethnography is an effective approach. Tourism enquiry should not only place an emphasis on people's experiences, but also place a more abstract focus on their ability to understand the world – the human consciousness.

References

Aas, C., Ladkin, A. and Fletcher, J. (2005) Stakeholder collaboration and heritage management. *Annals of Tourism Research* 32 (1), 28–48.

Abou-Shouk, M.A., Zoair, N., El-Barbary, M.N. and Hewedi, M.M. (2017) Sense of place relationship with tourist satisfaction and intentional revisit: Evidence from Egypt. *International Journal of Tourism Research* 20 (2), 172–181.

Academy of China Social Science (2022) *China Social Science Fund Project Guide 2020.* Beijing: Academy of China Social Science. http://fund.cssn.cn/skjj/

ACMHSDA (2003) Huanshan fengjingqu guangli weiyuanhui jianjie [A brief introduction to the Administrative Committee of Mount Huang Scenery Development Area]. Unpublished company document.

Adams, V. (1996) Karaoke as modern Lhasa, Tibet: Western encounters with cultural politics. *Cultural Anthropology* 11 (4), 510–546.

AFP News (2013) China boy's public urination peeves Hong Kongers. *AFP News Hong Kong*, 20 February. See http://sg.news.yahoo.com/china-boys-public-urination-peeves-hong-kongers-110418118.html (accessed 8 January 2014).

Agnew, J.A. (1987) *Place and Politics: The Geographical Mediation of State and Society.* London: Allen & Unwin.

Agnew, J.A. (2002) *Place and Politics in Modern Italy.* Chicago, IL: University of Chicago Press.

Agnew, J.A. (2011) Space and place. In J.A. Agnew and D.N. Livingstone (eds) *The Sage Handbook of Geographical Knowledge* (pp. 316–330). Los Angeles, CA: Sage.

Ahmed, Z.U., Ghingold, M. and Dahari, Z. (2007) Malaysian shopping mall behavior: An exploratory study. *Asia Pacific Journal of Marketing and Logistics* 19 (4), 331–348.

Allcock, J.B. (1995) International tourism and the appropriation of history in the Balkans. In M.F. Lanfant, J.B. Allcock and E.M. Bruner (eds) *International Tourism: Identity and Change* (pp. 101–112). London: Sage Publications.

Allen, J., O'Toole, W., Harris, R. and McDonnell, R.L. (2011) *Festival and Special Event Management.* Milton: Wiley.

Ames, H., Glenton, C. and Lewin, S. (2019) Purposive sampling in a qualitative evidence synthesis: A worked example from a synthesis on parental perceptions of vaccination communication. *BMC Medical Research Methodology* 19 (1), 19–26.

Amoiradis, C., Velissariou, E. and Stankova, M. (2021) Tourism as a social-cultural phenomenon: A critical analysis. *Journal of Social and Political Sciences* 4 (2), 10–21.

Anderson, B.R. (1983) *Imagined Communities: Reflections On the Origin and Spread of Nationalism.* London: Verso.

Andrade, C. (2020) The inconvenient truth about convenience and purposive samples. *Indian Journal of Psychological Medicine* 43 (1), 86–88.

Anuar, A.N.A., Ahmad, H., Jusoh, H. and Hussain, M.Y. (2012) Understanding the factors influencing formation of tourist friendly destination concept. *Journal of Management and Sustainability* 2 (1), 106–114.

Ap, J. (2001) Tourists perceptions of their level of contact with host residents. *Pacific Tourism Review* 5 (1), 51–58.

APM (2009) Introduction to Beijing APM, blog (in Chinese). See https://www.douban.com/event/10987584/discussion/18556428/ (accessed 15 December 2021).

Apple Daily (2000) A9, 26 July 2000.

Apple Daily (2010) Neidi tong haifangcheng dangzhong dabian [Mainland Chinese kid defecating in public in Harbour City]. *Apple Daily*, 27 September. See http://hk.apple.nextmedia.com/news/art/20100927/14492894 (accessed 8 January 2014).

Arensberg, C.M. and Niehoff, A.H. (1971) *Introducing Social Change: A Manual for Community Development*. Chicago, IL: Aldine Atherton, Inc.

Ashton, A. (2014) Drawing on the 'lived experience': An investigation of perception, ideation and praxis. *Art and Design Review* 2 (3), 46–61.

Ashworth, G.J. (1994) From history to heritage; from heritage to identity: In search of concepts and models. In G.J. Ashworth and P.J. Larkham (eds) *Building a New Heritage: Tourism, Culture and Identity in a New Europe* (pp. 13–30). London: Routledge.

Augé, M. (1995) *Non Places: Introduction to an Anthropology of Supermodernity*. London: Verso.

Aydede, M. (2019) The language of thought hypothesis. *Stanford Encyclopedia of Philosophy* (pp. 1–15). Stanford, CA: The Metaphysics Research Lab, Department of Philosophy, Stanford University. http://plato.stanford.edu/entries/language-thought/

Bach, J. (2002) The taste remains: Consumption, (n)ostalgia, and the production of East Germany. *Public Culture* 14 (2), 545–556.

Baker, F. (1993) The Berlin Wall: Production, preservation and consumption of a 20th century monument. *Antiquity* 67 (257), 709–734.

Baldacchino, G. (2004) The coming of age of island studies. *Tijdschrift voor economische en sociale geografie* [*Journal of Economic and Social Geography*] 95 (3), 272–283.

Baldacchino, G. (2008) Studying islands: On whose terms? Some epistemological and methodological challenges to the pursuit of island studies. *Island Studies Journal* 3 (1), 37–56.

Bao, J., Chen, G. and Ma, L. (2014) Tourism research in China: Insights from insiders. *Annals of Tourism Research* 45 (2014), 167–181.

Barboza, D. (2008) Beijing air terminal goes all out for the Games. *New York Times*, 2 May. See https://www.nytimes.com/2008/05/02/business/worldbusiness/02terminal.html

Barmé, G.R. (2000) *In the Red: On Contemporary Chinese Culture*. New York: Columbia University Press.

Beerli, A. and Martín, J.D. (2004) Factors influencing destination. *Annals of Tourism Research* 31 (3), 657–681.

Beijing Government (BMG) (2004) Beijing master planning. Unpublished planning document (in Chinese).

Beijing Statistics Bureau (BSB) (2000–2008) *Beijing Statistical Yearbook*. Beijing: China Statistics Press (in Chinese).

Bell, R.H. (1987) Conservation with a human face: Conflict and reconciliation in African land use planning. In D. Anderson and R. Grove (eds) *Conservation in African: People, Policies, and Practice* (pp. 79–101). Cambridge: Cambridge University Press.

Bernard, H.R. (2006) *Research Methods in Anthropology: Qualitative and Quantitative Approaches*. Lanham, MD: Alta Mira Press.

Billinge, M. (1977) In search of negativism: Phenomenology and historical geography. *Journal of Historical Geography* 3 (1), 55–67.

Birt, L., Scott, S., Cavers, D., Campbell, C. and Walter, F. (2016) Member checking: A tool to enhance trustworthiness or merely a nod to validation? *Qualitative Health Research* 26 (13), 1802–1811.
Blackstock, K. (2005) A critical look at community-based tourism. *Community Development Journal* 40 (1), 39–49.
Blunch, N.J. (2013) *Introduction to Structural Equation Modelling Using IBM SPSS Statistics and AMOS*. London: Sage Publications.
BMP (Cartographer) (2008) *Beijing Transportation and Tourist Map 2008*. Beijing: Beijing Map Press.
Bodley, J.H. (1990) *Victims of Progress*. Mountain View, CA: Mayfield Publishing Company.
Body, S. and Singh, S. (2003) Destination communities: Structures, resources and types. In S. Singh, D.J. Timothy and R.K. Dowling (eds) *Tourism in Destination Communities* (pp. 19–33). Oxford: CAB International.
Boniface, P. and Fowler, P.J. (1993) *Heritage and Tourism in the 'Global Village'*. London: Routledge.
Boorstin, D.J. (1964) *The Image: A Guide to Pseudo-Events in America*. New York: Harper & Row.
Bosman, C. and Dredge, D. (2011) Histories of Placemaking in the Gold Coast City: The Neoliberal Norm, the State Story and the Community Narrative. Griffith University Urban Research Program, Research Paper, 33. https://research-repository.griffith.edu.au/handle/10072/149863.
Bowitz, E. and Ibenholt, K. (2009) Economic impacts of cultural heritage: Research and perspectives. *Journal of Cultural Heritage* 10 (1), 1–8.
Boyd, S. (2002) Cultural and heritage tourism in Canada: Opportunities, principles and challenges. *Tourism and Hospitality Research* 3 (3), 211–233.
Boyd, S.W. and Butler, R.W. (eds) (2000) *Tourism and National Parks: Issues and Implications*. Chichester: Wiley.
Boym, S. (2001) *The Future of Nostalgia*. New York: Basic Books.
Brett, D. (1996) *The Construction of Heritage*. Cork: Cork University Press.
Bruner, E.M. (2005) *Culture on Tour: Ethnographies of Travel*. Chicago, IL: University of Chicago Press.
Bryden, J. (1973) *Tourism and Development: Case Study of Commonwealth Caribbean*. Cambridge: Cambridge University Press.
Burawoy, M. (2003) Revisits: An outline of a theory of reflexive ethnography. *American Sociological Review* 68 (5), 645–679.
Bureau for the Compilation and Translation of Works of Marx, Engels, Lenin and Stalin (1984) *Selected Works of Deng Xiaoping (1975–1982)*. Beijing: Foreign Languages Press.
Burgess, R.G. (1990) *In the Field: An Introduction to Field Research*. London: Routledge.
Butler, R.W. (1980) The concept of a tourism area cycle of evolution: Implications for the management of resources. *Canadian Geographer* 24 (1), 5–12.
Butler, R.W. (1993) Tourism: An evolutionary perspective. In J.G. Nelson, R.W. Butler and G. Wall (eds) *Tourism and Sustainable Development: Monitoring, Planning, Managing* (pp. 27–44). Waterloo: University of Waterloo, Heritage Resource Centre.
Butler, R. (1998) Sustainable tourism: Looking backwards in order to progress? In C.M. Hall and A.A. Lew (eds) *Sustainable Tourism: A Geographical Perspective* (pp. 25–34). London: Longman.
Butler, R.W. (2009) Tourism in the future: Cycles, waves or wheels? *Futures* 41 (6), 346–352.
Buttimer, A. (1976) Grasping the dynamism of the life-world. *Annals of the Association of American Geographers* 66 (2), 277–292.

Buttimer, A. and Seamon, D. (eds) (1980) *The Human Experience of Place and Space* (1st edn). New York: St. Martin's Press.

Byrne, B.M. (2010) *Structural Equation Modelling with AMOS: Basic Concepts, Applications, and Programming*. New York: Routledge.

Calhoun, C.J. (1997) *Nationalism*. Buckingham: Open University Press.

Campbell, R., Goodman-Williams, R., Feeney, H. and Fehler-Cabral, G. (2020) Assessing triangulation across methodologies, methods, and stakeholder groups: The joys, woes, and politics of interpreting convergent and divergent data. *American Journal of Evaluation* 41 (1), 125–144.

Campelo, A. and Aitken, R. (2011) Travelling to the past: Narratives of place and national identity on the Chatham Islands. In E. Frew and E. White (eds) *Tourism and National Identities: An International Perspective* (pp. 191–201). Abingdon: Routledge.

Cao, D.S. and Hu, M.D. (2020) Anti-poverty efforts earn everyone's praise. *China Daily*. See http://www.chinadaily.com.cn/a/202010/17/WS5f8a2920a31024ad0ba7f3e9.html

Cao, J., Zhang, J., Wang, C., Hu, H. and Yu, P. (2019) How far is the ideal destination? Distance desire, ways to explore the antinomy of distance effects in tourist destination choice. *Journal of Travel Research* 59 (2), 1–17.

Caraba, C.C. (2011) Communist heritage tourism and red tourism: Concepts, development and problems. *Cinq Continents* 1 (1), 29–39.

Carlsen, J. and Taylor, A. (2003) Mega-events and urban renewal: The case of the Manchester 2002 commonwealth games. *Event Management* 8 (1), 15–22.

Carlsen, J. and Zulfa, M. (2016) Island tourism competitiveness and sustainability in the Maldives. In P. Modica and M. Uysal (eds) *Sustainable Island Tourism: Competitiveness and Quality of Life* (pp. 75–92). Wallingford: CABI.

Castree, N. (2003) Place: Connections and boundaries in an interdependent world. In S.L. Holloway, S.P. Rice and G. Valentine (eds) *Key Concepts in Geography* (pp. 165–185). London: Sage.

Castree, N. (2009) Place: Connections and boundaries in an interdependent world. In N. Clifford, S. Holloway, S.P. Rice and G. Valentine (eds) *Key Concepts in Geography* (pp. 153–172). London: Sage.

Caton, K. and Santos, C.A. (2007) Closing the hermeneutic circle: Photographic encounters with the other. *Annals of Tourism Research* 35 (1), 7–26.

Cave, J. and Baldacchino, G. (2012) Come visit, but don't overstay: Critiquing a welcoming society. *International Journal of Culture, Tourism and Hospitality Research* 6 (2), 145–153.

Cave, J. and Brown, K.G. (2012) Island tourism destinations: An editorial introduction to the special issue. *International Journal of Culture, Tourism and Hospitality Research* 6 (2), 95–113.

Chalise, K.R. (2019) Lyotard's 'The Postmodern Condition': Issues of knowledge and legitimacy. *Literary Endeavour* 10 (4), 122–129.

Chamberlain, H.B. (1994) Coming to terms with civil society. *The Australian Journal of Chinese Affairs* 31 (January), 113–117.

Chambers, R. (1991) Rapid (or relaxed) and participatory rural appraisal: Notes on practical approaches and methods. *Qualitative Research Methods Newsletter* 2, 11–16.

Chang, X.M. and Wang, K.Y. (eds) (2020) Zhonggong zhongyang guanyu xinshidai tuijin xibu dakaifa xingcheng xin geju de zhidaoyijian [Guiding opinions of the Chinese Communist Party Central Committee and the State Council on promoting the western development in the new era and forming a new pattern]. *People's Net*. See http://cpc.people.com.cn/n1/2020/0518/c419242-31712395.html (accessed 6 October 2022).

Cheer, J. (2019) Tourism on small islands: The urgency for sustainability indicators. In J. Randall (ed.) *The 21st Century Maritime Silk Road Islands Economic Cooperation Forum: Annual Report on Global Islands* (pp. 129–152). Charlottetown: Island Studies Press.

Chen, J. (2008) *Nation, Ethnicity and Cultural Strategies: Three Waves of Ethnic Representation in post-1949 China*. New Brunswick, NJ: Rutgers University.
Chen, B. and Yu, L. (1992) *Strategy for Tourism Development in Shenzhen*. Shanghai: Tongji University Press.
China National State Council (1981) *Zongjiao huodong changsuo guanli tiaoli* [*Regulations Regarding Management and Operation of Religious Sites and Temples*]. Beijing: China National State Council.
China National State Council. (1998) *Law of the People's Republic of China on land management, the National People's Congress (promulgated 29 August 1998 as Order No. 8 of the President of the People's Republic of China*. Beijing: China National State Council.
CNTA (1993) *The Yearbook of China Tourism Statistics*. Beijing: CNTA.
CNTA (1994) China Folklore 1995 (printed tourist brochure). Beijing: China National Tourism Administration.
CNTA (1999) *The Yearbook of China Tourism Statistics*. Beijing: CNTA.
CNTA (2006) *The Yearbook of China Tourism Statistics 2006*. Beijing: China Travel and Tourism Press.
CNTA (2009) *The Yearbook of China Tourism Statistics 2009*. Beijing: China Travel and Tourism Press.
CNTA (2015) *Lvyou jingqu zhiliang dengji pingding guanli banfa* [*Quality Grade Evaluation and Division of Scenic Spots*]. Beijing: China National Tourism Administration.
Chhabra, D. (2009) Proposing a sustainable marketing framework for heritage tourism. *Journal of Sustainable Tourism* 17 (3), 303–320.
Chhabra, D., Healy, R. and Sills, E. (2003) Staged authenticity and heritage tourism. *Annals of Tourism Research* 30 (3), 702–719.
China Travel Service HK Ltd (2000) *Hotel Reservations*. Hong Kong: China Travel Service HK Ltd.
China Tourism Academy (2018) *2017 nian zhongguo lvyouye tongji gongbao [Annual statistics report of China tourism industry 2017]*. Retrieved from http://www.ctaweb.org/html/2018-12/2018-12-28-15-55-12622.html
Chistyakova, O.V. (2020) Ethnic tourism as knowing other. *RUDN Journal of Philosophy* 24 (4), 720–729.
Cifçi, T. and Köybaşı, F. (2017) Geographic consciousness in sustainable education: Students' views. *Journal of Education and Practice* 8 (17), 49–55.
Choi, E.K. and Zhou, K.X. (2001) Entrepreneurs and politics in the Chinese traditional economy: Political connections and rent-seeking. *The China Review* 1, 111–136.
Choi, H.C. and Sirakaya, E. (2006) Sustainability indicators for managing community tourism. *Tourism Management* 27 (6), 1274–1289.
Choksi, N. (2012) Surface politics: Scaling multiscriptality in an Indian village market. *Journal of Linguistic Anthropology* 25 (1), 1–23.
Chow, V. (2014) Tradition or tourist spectacle? Safeguarding Hong Kong's culture comes at a price. Safeguarding the Cheung Chau Bun Festival and other cherished traditions comes with a price. *South China Morning Post*. See https://www.scmp.com/news/hong-kong/article/1587371/popularity-cheung-chau-bun-festival-shows-price-be-paid-preserving (accessed 2 October 2021).
Chronis, A. (2005) Constructing heritage at the Gettysburg storyscape. *Annals of Tourism Research* 32 (2), 386–406.
Churchill Jr, G.A. (1979) A paradigm for developing better measures of marketing constructs. *Journal of Marketing Research* 16 (1), 64–73.
Clarkson, B.E. (1995) A stakeholder framework for analysing and evaluating corporate social performance. *The Academy of Management Review* 20 (1), 92–117.
Cohen, E. (1979) A phenomenology of tourist experiences. *The Journal of the British Sociological Association* 13 (2), 179–201.

Cohen, E. (1989) Primitive and remote: Hill tribe trekking in Thailand. *Annals of Tourism Research* 16 (1), 30–61.
Cohen, E. (1996) *Thai Tourism: Hilltribes, Islands and Tourist-Oriented Prostitution*. Bangkok: White Lotus.
Colaizzi, P.F. (1978) Psychological research as a phenomenologist views it. In R.S. Valle and M. King (eds) *Existential Phenomenological Alternatives for Psychology* (pp. 48–71). New York: Oxford University Press.
Coles, T. (2003a) The emergent tourism industry in Eastern Germany a decade after unification. *Tourism Management* 24 (2), 217–226.
Coles, T. (2003b) Urban tourism, place promotion and economic restructuring: The case of post-socialist Leipzig. *Tourism Geographies* 5 (2), 190–219.
Colombino, A. (2009) Multiculturalism and time in Trieste: Place-marketing images and residents' perceptions of a multicultural city. *Social and Cultural Geography* 10 (3), 279–297.
Comerio, N. and Strozzi, F. (2018) Tourism and its economic impact: A literature review using bibliometric tools. *Tourism Economics* 25 (1), 109–131.
Concu, N. and Atzeni, G. (2012) Conflicting preferences among tourists and residents. *Tourism Management* 33 (6), 1293–1300.
Consumer News and Business Channel (CNBC) (2022) China economy: WHO chief says China's zero-Covid policy is not 'sustainable'. *Reuters*. See https://www.cnbc.com/2022/05/11/who-chief-says-chinas-zero-covid-policy-is-not-sustainable.html (accessed June 2022).
Cook, I. and Crang, M. (1995) *Doing Ethnographies*. Norwich: University of East Anglia Press.
Corsane, G., Davis, P. and Mutas, D. (2009) Place, local distinctiveness and local identity: Ecomuseum approaches in Europe and Asia. In M. Anico and E. Peralta (eds) *Heritage and Identity: Engagement and Demission in the Contemporary World* (pp. 47–62). Abingdon: Routledge.
Craik, J. (1991) *Resorting to Tourism: Cultural Policies for Tourist Development in Australia*. Sydney: Allen & Unwin.
Crang, M. and Thrift, N.J. (2000) *Thinking Space*. New York: Routledge.
Cresswell, T. (1996) *In Place/Out of Place: Geography, Ideology, and Transgression*. Minneapolis, MN: University of Minnesota Press.
Cresswell, T. (1997) Weeds, plagues, and bodily secretions: A geographical interpretation of metaphors of displacement. *Annals of the Association of American Geographers* 87 (2), 330–345.
Cresswell, T. (2007) *Place: A Short Introduction*. Malden, MA: Blackwell.
Cresswell, T. (2010) Towards a politics of mobility. *Environment and Planning D: Society and Space* 28 (1), 17–31.
Cresswell, T. (2014) *Place: A Short Introduction*. Hoboken, NJ: John Wiley & Sons.
Croes, R. and Semrad, K.J. (2015) The relevance of cultural tourism as the next frontier for small island destinations. *Journal of Hospitality and Tourism Research* 39 (4), 469–491.
Cui, Q., Liao, X.H. and Xu, H.G. (2017) Tourist experience of nature in contemporary China: A cultural divergence approach. *Journal of Tourism and Cultural Change* 15 (3), 248–264.
Culler, J. (1981) Semiotics of tourism. *The American Journal of Semiotics* 1 (1/2), 127–140.
Dann, G.M.S. (1993) Limitations in the use of 'nationality' and 'country of residence' variables. In D.G. Pearce and R.W. Butler (eds) *Tourism Research: Critiques and Challenges* (pp. 88–112). London: Routledge.
Dann, G. and Cohen, E. (1996) Sociology and tourism. In Y. Apostopoloulos, S. Leivadi and A. Yiannakis (eds) *The Sociology of Tourism: Theoretical and Empirical Investigation* (pp. 301–314). London: Routledge.

Davis, C.A. (2007) *Reflexive Ethnography: A Guide to Researching Selves and Others*. Abingdon: Routledge.

Davis, F. (1979) *Yearning for Yesterday: A Sociology of Nostalgia*. New York: Free Press.

Dawson, A.H. (1999) From glittering icon to... *The Geographical Journal* 165 (2), 154–160.

de Jong, S. and Dannecker, P. (2018) Connecting and confronting transnationalism: Bridging concepts and moving critique. *Identities* 25 (5), 493–506.

Decrop, A. (1999) Triangulation in qualitative tourism research. *Tourism Management* 20 (1), 157–161.

Desbois, H. (2016) The map and the text: A geographic reading of 'Wandering Rocks' (Ulysses, by Joyce, Chapter 10). *L'Espace Géographique* 45 (4), 355–367.

Dickson, C.N., Lawan, A. and Godwin, O. (2020) Issues in development and underdevelopment: A theoretical discourse. *International Journal of Academic Multidisciplinary Research* 4 (8), 147–154.

Dolezal, C. and Novelli, M. (2022) Power in community-based tourism: Empowerment and partnership in Bali. *Journal of Sustainable Tourism* 30 (10), 2352–2370.

Douglas, J.D. (1985) *Creative Interviewing*. Beverly Hills, CA: Sage.

Dujisin, Z. (2007) *Routledge Handbook of Tourism Cities*. Abingdon: Routledge.

Dutta, M., Banerjee, S. and Husain, Z. (2007) Untapped demand for heritage: A contingent valuation study of Prinsep Ghat, Calcutta. *Tourism Management* 28 (1), 83–95.

Dwyer, L., Mellor, R., Mistilis, N. and Mules, T. (2000) A framework for assessing 'tangible' and 'intangible' impacts of events and conventions. *Event Management* 6 (3), 175–189.

Echtner, C.M. and Ritchie, J.R. (1991) The meaning and measurement of destination image. *The Journal of Tourism Studies* 2 (2), 2–12.

Edensor, T. (1997) National identity and the politics of memory: Remembering Bruce and Wallace in symbolic space. *Environment and Planning D: Society & Space* 15 (2), 175–194.

Edensor, T. (2009) Tourism and performance. In T. Jamal and M. Robinson (eds) *The Sage Handbook of Tourism* (pp. 543–557). London: Sage.

Edmonds, R.L. (1999) China's environmental problems. In R.E. Gamer (ed.) *Understanding Contemporary China* (pp. 237–266). Boulder, CO: Lynne Rienner Publishers.

Edwards, F. (1988) *Environmentally Sound Tourism Development in the Caribbeans*. Calgary: Calgary Press.

Ekici, A., Sirgy, M.J., Lee, D., Yu, G.B. and Bosnjak, M. (2018) The effects of shopping well-being and shopping ill-being on consumer life satisfaction. *Applied Research Quality Life* 13 (2018), 333–353.

Emerson, R.M., Fretz, R.I. and Shaw, L.L. (1995) *Writing Ethnographic Fieldnotes*. Chicago, IL: University of Chicago Press.

Erkip, F. (2003) The shopping mall as an emergent public space in Turkey. *Environment and Planning A* 35 (6), 1073–1093.

Essex, S. and Chalkley, B. (2004) Mega-sporting events in urban and regional policy: A history of the Winter Olympics. *Planning Perspectives* 19 (2), 201–204.

Evans-Pritchard, D. (1993) Ancient art in modern context. *Annals of Tourism Research* 20 (1), 9–31.

Falk, J.H., Ballantyne, R., Packer, J. and Benckendorff, P. (2012) Travel and learning: A neglected tourism research area. *Annals of Tourism Research* 39 (2), 908–927.

Fallon, L.D. and Kriwoken, L.K. (2003) Community involvement in tourism infrastructure: The case of the Strahan Visitor Centre, Tasmania. *Tourism Management* 24 (3), 289–308.

Farquhar, J. and Zhang, Q. (2005) Biopolitical Beijing: Pleasure, sovereignty, and self-cultivation in China's capital. *Cultural Anthropology* 20 (3), 303–327.

Ferrara, G. (2017) The paradox of tourism in underdeveloped economies. *Journal of Economics and Finance* 8 (3), 40–43.

Figueroa-Domecq, C., de Jong, A. and Williams, A.M. (2020) Gender, tourism & entrepreneurship: A critical review. *Annals of Tourism Research* 84, 102980. https://doi.org/10.1016/j.annals.2020.102980.

Finkelman, R.B. and Centeno, J.A. (2020) Guizhou Province China: The birthplace of modern medical geology. *Acta Geochimica* 39 (1), 155–159.

Fletcher, J., Fyall, A., Gilbert, D. and Wahill, S. (2018) *Tourism: Principles and Practice*. London: Pearson Education.

Foote, K.E. and Azaryahu, M. (2009) Sense of place. In K. Kitchin and N. Thrift (eds) *International Encyclopedia of Human Geography* (pp. 96–100). Amsterdam: Elsevier.

Fornell, C. and Larcker, D.F. (1981) Structural equation models with unobservable variables and measurement error: Algebra and statistics. *Journal of Marketing Research* 18 (3), 382–388.

Foucault, M. (1980) *Power/Knowledge: Selected Interviews and Other Writings, 1972–1977*. New York: Pantheon Books.

Foucault, M. (1990) *The History of Sexuality Volume 1: An Introduction*. New York: Vintage.

Fraser, R. (2020) Cultural heritage, ethnic tourism, and minority-state relations amongst the Orochen in north-east China. *International Journal of Heritage Studies* 26 (2), 178–200.

Freestone, R. and Liu, E. (2016) *Place and Placelessness Revisited*. Abingdon: Routledge.

Frigg, R. and Hartmann, S. (2006) Models in science. In *Stanford Encyclopedia of Philosophy* (pp. 1–23). Stanford, CA: The Metaphysics Research Lab, Department of Philosophy, Stanford University.

Fuller, A.M., Bao, J.G., Liu, Y. and Zhou, X.Y. (2022) Establishing tourism sustainability in a globally important agricultural heritage system in China: A case of social and eco-system recovery. *Growth and Change* 53 (3), 1267–1281. https://doi.org/10.1111/grow.12605.

Furnham, A. (1984) Tourism and culture shock. *Annals of Tourism Research* 11 (1), 41–58.

Gallarza, M.G., Saura, I.G. and Garcia, H.C. (2002) Destination image: Towards a conceptual framework. *Annals of Tourism Research* 29 (1), 56–78.

Geertz, C. (1973) *The Interpretation of Cultures: Selected Essays*. New York: Basic Books.

George, E.W., Mair, H. and Reid, D.C. (2009) *Rural Tourism Development: Localism and Cultural Change*. Bristol: Channel View Publications.

Getz, D. (1997) *Event Management & Event Tourism*. New York: Cognizant Communication Corp.

Getz, D. (2008) Event tourism: Definition, evolution, and research. *Tourism Management* 29 (3), 403–428.

Giddens, A. (1984) *The Constitution of Society*. Cambridge: Polity Press.

Giddens, A. (1990) *The Consequences of Modernity*. Stanford, CA: Stanford University Press.

Giere, R.N. (2004) How models are used to represent reality. *Philosophy of Science* 71 (5), 742–752.

Gieryn, T.F. (2000) A space for place in sociology. *Annual Review of Sociology* 26 (1), 463–496.

Giles, M., Giles, H. and Bernhold, Q. (2019) Cultural festivals as intergroup settings: A case study of Pacific islander identification. *Journal of Multilingual and Multicultural Development* 40 (9), 1–15.

Giorgi, A. (2009) *The Descriptive Phenomenological Method in Psychology: A Modified Husserlian Approach*. Pittsburgh, PA: Duquesne University Press.

Goeldner, C. and Brent Ritchie, J.R. (2011) *Tourism: Principles, Practices, Philosophies* (12th edn). New York: John Wiley and Sons, Inc.

Google (2021) Google map. See https://www.researchgate.net/figure/Location-map-of-Shenzhen-and-Hong-Kong_fig28_264534141 (accessed 16 November 2021).
Goorha, P. (2010) Modernisation theory. In R.A. Denemark (ed.) *The International Studies Encyclopedia* (pp. 1–26). Oxford: Blackwell Publishing.
Goulding, C. and Domic, D. (2009) Heritage, identity and ideological manipulation: The case of Croatia. *Annals of Tourism Research* 36 (1), 85–102.
Göymen, K. (2000) Tourism and governance in Turkey. *Annals of Tourism Research* 27 (4), 1025–1048.
Graburn, N. (1989) Tourism: The sacred journey. In V. Smith (ed.) *Hosts and Guests: The Anthropology of Tourism* (pp. 21–36). Philadelphia, PA: University of Philadelphia Press.
Graburn, N. (2008) Reconstructing tradition: Tourism and modernity in China and Japan. *Revista Brasileira de Ciências Sociais* 23 (68), 11–21.
Granovetter, M. (2001) Economic action and social structure: The problem of embeddedness. In R. Swedberg and M. Granovetter (eds) *The Sociology of Economic Life* (pp. 51–76). Boulder, CO: Westview Press.
Gratton, C., Shibli, S. and Coleman, R. (2005) Sport and economic regeneration in cities. *Urban Studies* 42 (5–6), 985–999.
Greenhalgh, T. and Stones, R. (2010) Theorising big IT programmes in healthcare: Strong structuration theory meets actor-network theory. *Social Science and Medicine* 70 (9), 1285–1294.
Greenwood, D.J. (1982) Cultural authenticity. *Cultural Survival Quarterly* 6 (3), 27–28.
Greenwood, D.J. (1989) Culture by the pound: An anthropological perspective on tourism as cultural commodification. In V.L. Smith (ed.) *Hosts and Guests: The Anthropology of Tourism* (pp. 176–186). Philadelphia, PA: University of Pennsylvania Press.
Griffiths, I. and Sharpley, R. (2012) Influences of nationalism on tourist–host relationships. *Annals of Tourism Research* 39 (4), 2051–2072.
Grydehøj, A. (2017) A future of island studies. *Island Studies Journal* 12 (1), 3–1.
Grydehøj, A. (2020) Critical approaches to island geography. *Area* 52 (1), 2–5.
Gunn, C.A. (1988) *Vacationscape: Designing Tourist Regions* (2nd edn). New York: Van Nostrand Reinhold.
Gunn, C.A. (2002) *Tourism Planning: Basics, Concepts, Cases*. New York: Routledge.
Guo, R. (2021) China approves arrest of former Hainan official on corruption charges. *South China Morning Post*, 28 January. See https://www.scmp.com/news/china/politics/article/3119675/china-approves-arrest-former-hainan-official-corruption-charges (accessed 21 February 2021).
Guo, Y. (2004) *Cultural Nationalism in Contemporary China: The Search for National Identity Under Reform*. London: Routledge.
Gursoy, D. and Rutherford, D.G. (2004) Host attitudes toward tourism: An improved structural model. *Annals of Tourism Research* 31 (3), 495–516.
Gursoy, D., Jurowski, C. and Uysal, M. (2002) Resident attitudes: A structural modeling approach. *Annals of Tourism Research* 29 (1), 79–105.
Hainan Tourism Bureau (2003) Tourism statistics 1995–2001 (unpublished government documents). Hainan Provincial Government, Haikou, Hainan, China.
Hair, J.F., Black, W.C. and Babin, B.J. (2010a) *Multivariate Data Analysis: A Global Perspective* (7th edn). New York: Pearson Education.
Hair, J.F., Anderson, R.F., Tatham, R.L. and Black, W.C. (2010b) *Multivariate Data Analysis*. Upper Saddle River, NJ: Prentice Hall/Pearson.
Hall, C.M. (1992) *Hallmark Tourist Events: Impacts, Management, and Planning*. London: Belhaven Press.
Hall, C.M. (1994) *Tourism and Politics: Policy, Power and Place*. Chichester: Wiley.
Hall, C.M. (2005) *Tourism: Rethinking the Social Science of Mobility*. Harlow: Pearson/Prentice Hall.

Hamilton-Smith, E. (1987) Four kinds of tourism. *Annals of Tourism Research* 14 (3), 332–334.

Hammersley, M. and Atkinson, P. (2019) *Ethnography: Principles in Practice*. Abingdon: Routledge.

Hammitt, W.E., Backlund, E.A. and Bixler, R.D. (2006) Place bonding for recreation places: Conceptual and empirical development. *Leisure Studies* 25 (1), 17–41.

Han, M. (2001) Mao tourism of Shaoshan. In C.B. Tan, S.C.H. Cheung and H. Yang (eds) *Tourism, Anthropology and China* (pp. 216–222). Bangkok: White Lotus.

Harrison, D. (1988) *The Sociology of Modernisation and Development*. London: Unwin Hyma.

Harrison, D. (1992) International tourism and the less developed countries: The background. In D. Harrison (ed.) *International Tourism and the Less Developed Countries* (pp. 1–17). London: Belhaven Press.

Harvey, D. (1989) *The Condition of Postmodernity: An Enquiry into the Origins of Cultural Change*. Oxford: Blackwell.

Harvey, D. (1996) *Justice, Nature and the Geography of Difference*. Cambridge, MA: Blackwell.

Hatton, M.J. (1999) *Community-Based Tourism in the Asia-Pacific*. Vancouver: APEC Publication.

Hay, P. (2006) A phenomenology of islands. *Island Studies Journal* 1 (1), 19–42.

Hay, P. (2013) What the sea portends: A reconsideration of contested island tropes. *Island Studies Journal* 8 (2), 209–232.

Hechter, M. (2000) *Containing Nationalism*. Oxford: Oxford University Press.

Henderson, J.C. (2007) Communism, heritage and tourism in East Asia. *International Journal of Heritage Studies* 13 (3), 240–254.

Henderson, K.A. (1991) *Dimensions of Choice: A Qualitative Approach to Recreation, Parks, and Leisure Research*. State College, PA: Venture Publishing, Inc.

Hersh, P. (2001) Beijing savours Olympic victory. *Chicago Tribune*, 14 July. See https://www.chicagotribune.com/news/ct-xpm-2001-07-14-0107140091-story.html (accessed 16 November 2021).

Hibberts, M., Johnson, R.B. and Hudson, K. (2012) Common survey sampling techniques. In L. Gideon (ed.) *Handbook of Survey Methodology for the Social Sciences* (pp. 53–74). New York: Springer.

Hibbort, P. (1990) Cook's Peking home. *Time Traveller* 15, 8–9.

Higham, J. and Miller, G. (2018) Transforming societies and transforming tourism: Sustainable tourism in times of change. *Journal of Sustainable Tourism* 26 (1), 1–8.

Hinch, T.D. and Li, Y. (1994) Developing tourism in China: Opportunities and challenges. In F. Go and H. Qu (eds) *Reducing Barriers to International Tourism* (pp. 80–87). Hong Kong: Hong Kong Polytechnic University Press.

Hitchcock, M. (1999) Tourism and ethnicity: Situational perspectives. *International Journal of Tourism Research* 1 (1999), 17–32.

Hoerder, D. (2010) *To Know Our Many Selves: From the Study of Canada to Canadian Studies*. Edmonton: Athabasca University Press.

Hollingworth, J.R. and Boyer, R. (eds) (1997) *Contemporary Capitalism: The Embeddedness of Institutions*. Cambridge: Cambridge University Press.

Hollinshead, K. and Hou, C.X. (2012) The seductions of 'soft power': The call for multifronted research into the articulative reach of tourism in China. *Journal of China Tourism Research* 8 (7), 227–247.

Hollinshead, K., Ateljevic, I. and Ali, N. (2009) Worldmaking agency – worldmaking authority: The sovereign constitutive role of tourism. *Tourism Geographies* 11 (4), 427–433.

Hong, G. (2017) Locating Zhuhai between land and sea: A relational production of Zhuhai, China, as an island city. *Island Studies Journal* 12 (2), 7–24.

Hong Kong Tourism Board (HKTB) (2013a) A Statistical Review of Hong Kong Tourism 2012, government post. See http://securepartnernet.hktb.com/filemanager/intranet/dept_info/private_20/paper/Stat-Review/Stat_Review2012/Stat_Review_2012_1.pdf (accessed 3 January 2014).

HKTB (2013b) Press release. Fifth American Express Hong Kong Wine and Dine Festival opens today at new central harbourfront. Hong Kong Tourism Board. Released on 31 October 2013.

HKTB (2014) Press release: CCB (Asia) Hong Kong Wine and Dine Festival Opening Party draws to a close, 180,000 participants recorded in four-day event. Hong Kong Tourism Board. Released on 31 December 2014.

HKTB (2019) *A Statistical Review of Hong Kong Tourism 2019: Insights and Research.* Hong Kong: Hong Kong Tourism Board.

Hong, L.P., Ngo, H.T. and Pham, L.T. (2021) Community-based tourism: Opportunities and challenges: A case study in Thanh Ha pottery village, Hoi An city, Vietnam. *Cogent Social Sciences* 7(1). https://doi.org/10.1080/23311886.2021.1926100

Horn, C. and Simmons, D. (2002) Community adaptation to tourism: Comparisons between Rotorua and Kaikoura, New Zealand. *Tourism Management* 23 (1), 133–143.

Horrigan-Kelly, M., Millar, M. and Dowling, M. (2016) Understanding the key tenets of Heidegger's philosophy for interpretive phenomenological research. *International Journal of Qualitative Methods* 15 (2016), 1–8.

Howe, B.L. and Jackson, J.J. (1985) *Teaching Effectiveness Research.* Victoria BC: University of Victoria Press.

Hu, Z.M. (2005) Developing red tourism energetically: Carrying on education of patriotism and revolutionary tradition (in Chinese). In Jiang Xi Provincial Government (ed.) *Proceedings of National Red Tourism Conference* (pp. 2–9). Nanchang, China.

Hunter, J.E. (2020) Towards a cultural analysis: The need for ethnography in interpretation research. *Journal of Interpretation Research* 17 (2), 47–58.

Inskeep, E. (1991) *Tourism Planning: An Integrated and Sustainable Development Approach.* New York: Van Nostrand Reinhold.

Ip, R. (2012) Root of Hong Kong's anti-mainland anger. *South China Morning Post*, 14 October. See http://www.scmp.com/comment/insight-opinion/article/1060412/root-hong-kongs-anti-mainland-anger?page=all (accessed 6 January 2013).

Ishii, K. (2012) The impact of ethnic tourism on hill tribes in Thailand. *Annals of Tourism Research* 39 (1), 290–310.

Islam, M. (2020) Reflection note: Confessions of a Muslim researcher – considering identity in research. *International Journal of Social Research Methodology* 23 (5), 509–513.

Ivanova, M., Buda, D. and Burrai, E. (2020) Creative and disruptive methodologies in tourism studies. *Tourism Geographies* 23 (1–2), 1–10.

Iyer, P. (1988) *Video Night in Kathmandu and Other Reports from the Not-so-Far East.* New York: Knopf.

Jackson, E.L. (1991) Shopping and leisure: Implications of West Edmonton Mall for leisure and for leisure research. *Canadian Geographer/Le Géographe canadien* 35 (3), 280–287.

Jacobson, M. and Larsen, S.C. (2014) Ethnographic fiction for writing and research in cultural geography. *Journal of Cultural Geography* 31 (2), 179–193.

Jago, L., Dwyer, L., Lipman, G., van Lill, D. and Vorster, S. (2010) Optimising the potential of mega-events: An overview. *International Journal of Event and Festival Management* 1 (3), 220–237.

Jamal, T. and Hollinshead, K. (2001) Tourism and the forbidden zone: The under-served power of qualitative enquiry. *Tourism Management* 22 (1), 63–82.

Jayant, L. and Fahimul, Q. (2004) Introduction: Democracy and development in Asia in the 21st century: In search of popular democratic alternatives. In L. Jayant and Q. Fahimul (eds) *Democracy and Civil Society in Asia Volume 2: Democratic Transitions and Social Movements in Asia* (pp. 1–31). Palgrave: Macmillan.

Jensen, O.B. (2014) *Staging Mobilities*. New York: Routledge.

Johnson, N.C. (1995) Cast in stone: Monuments, geography, and nationalism. *Environment and Planning D: Society & Space* 13 (1), 51–65.

Johnson, N.C. (1999) Framing the past: Time, space and the politics of heritage tourism in Ireland. *Political Geography* 18 (2), 187–207.

Johnston, R.J., Gregory, D. and Smith, D.M. (1988) *The Dictionary of Human Geography*. Oxford: Blackwell.

Johnston, R.J., Gregory, D., Pratt, G. and Watts, M.J. (eds) (2000) *The Dictionary of Human Geography*. Malden, MA: Blackwell Publishers.

Jorgensen, B.S. and Stedman, R.C. (2001) Sense of place as an attitude: Lakeshore owners' attitudes toward their properties. *Journal of Environmental Psychology* 21 (30), 233–248.

Jorgensen, B.S. and Stedman, R.C. (2006) A comparative analysis of predictors of sense of place dimensions: Attachment to, dependence on, and identification with lakeshore properties. *Journal of Environmental Management* 79 (3), 316–327.

Kaam, A.V. (1966) *Existential Foundations of Psychology*. Pittsburgh, PA: Duquesne University Press.

Kalinowski, K.M. (1992) Case study – universities and educational travel program: The University of Alberta. In B. Weiler and C.M. Hall (eds) *Special Interest Tourism* (pp. 27–36). London: Belhaven Press.

Kalinowski, K.M. and Weiler, B. (1992) Review: Educational travel. In B. Weiler and C.M. Hall (eds) *Special Interest Tourism* (pp. 15–26). London: Belhaven Press.

Kallen, J. (2009) Tourism and representation in the Irish linguistic landscape. In E. Shohamy and D. Gorter (eds) *Linguistic Landscape: Expanding the Scenery* (pp. 270–284). London: Routledge.

Ke, L., Zhang, J. and Li, Q. (2010) Shufa jingguan lvyoudi youke ganzhi yixiang yingxiangyinzi fenxi: Yi Huilin Diecaishan jingqu weili [Analyzing the influencing factors of tourists' perceived image in calligraphy scenic spots: A case study of Diecai Mountain Scenic Spot in Guilin, China]. *Zhong Guo Rong Yan* [*China Karst*] 29 (1), 87–92.

Kearns, R., Collins, D. and Bates, L. (2017) 'It's freedom!': Examining the motivations and experiences of coastal freedom campers in New Zealand. *Leisure Studies* 36 (3), 395–408.

Keller, C.P. (1987) Stages of peripheral tourism development: Canada's Northwest Territories. *Tourism Management* 8 (1), 20–32.

Keogh, B. (1990) Public participation in community tourism planning. *Annals of Tourism Research* 17 (3), 449–465.

Kianicka, S., Buchecker, M., Hunziker, M. and Müller-Böker, U. (2006) Locals' and tourists' sense of place: A case study of a Swiss Alpine village. *Mountain Research and Development* 26 (1), 55–63.

Kim, H. and Jamal, T. (2007) Touristic quest for existential authenticity. *Annals of Tourism Research* 34 (1), 181–201.

Kim, S.S., Timothy, D.J. and Han, H.C. (2007) Tourism and political ideologies: A case of tourism in North Korea. *Tourism Management* 28 (4), 1031–1043.

Kindon, S.a.A.L. (2002) From mitigation to negotiation: Ethics and the geographical imagination in Aotearoa/New Zealand. *New Zealand Geographer* 58 (1), 14–22.

Klieger, P.C. (1990) Close encounters: Intimate tourism in Tibet. *Cultural Survival Quarterly* 14 (2), 38–42.

Knox, P.L. (2005) Creating ordinary places: Slow cities in a fast world. *Journal of Urban Design* 10 (1), 1–11.

Ko, V. (2012) Trouble down south: Why Hong Kong and mainland Chinese aren't getting along. *Time*, 24 January. See http://world.time.com/2012/01/24/trouble-down-south-why-hong-kong-and-mainland-chinese-arent-getting-along/ (accessed 7 January 2014).

Koelsch, W.A. (2004) Squinting back at Strabo. *Geographical Review* 94 (4), 502–518.

Kvale, S. (1996) *Interviews: An Introduction to Qualitative Research Interviewing*. Beverly Hills, CA: Sage.

Kwong, C. and Li, Y.P. (2020) In-place or out-of-place? Host–guest encounter under 'One Country, Two Systems'. *Current Issues in Tourism* 23 (12), 1460–1479.

Kyle, G., Graefe, A., Manning, R. and Bacon, J. (2004) Effect of activity involvement and place attachment on recreationists' perceptions of setting density. *Journal of Leisure Research* 36 (2), 209–231.

Kylie, A. (2017) Connecting to 6,000 years of history at Wanuskewin Heritage Park. *Canadian Geographic*. See https://www.canadiangeographic.ca/article/connecting-6000-years-history-wanuskewin-heritage-park (accessed 24 January 2022).

Lai, K. and Li, Y. (2012) Core-periphery structure of destination image: Concept, evidence, and implication. *Annals of Tourism Research* 39 (3), 1359–1379.

Lai, K., Li, Y.P. and Feng, X. (2006) Gap between tourism planning and implementation: A case of China. *Tourism Management* 27 (6), 1171–1180.

Lambert, V., Glacken, M. and McCarron, M. (2011) Employing an ethnographic approach: Key characteristics. *Nurse Researcher* 19 (1), 17–24.

Landry, R. and Bourhis, R. (1997) Linguistic landscape and ethnolinguistic vitality: An empirical study. *Journal of Language and Social Psychology* 16 (1), 23–49.

Lanza, E. and Woldemariam, H. (2014) Multilingualism and local literacy practices in Ethiopia: Language contact in regulated and unregulated spaces. *Multilingual Margins* 1 (1), 74–100.

Larsen, J. (2014) (Auto)Ethnography and cycling. *International Journal of Social Research Methodology* 17 (1), 59–71.

Larsen, S.C. and Johnson, J.T. (2011) Toward an open sense of place: Phenomenology, affinity, and the question of being. *Annals of the Association of American Geographers* 102 (3), 32–46.

Larson, J.E. (2009) *Educational Psychology: Cognition and Learning, Individual Differences and Motivation*. New York: Nova Science Publishers.

Latham, A. (2003) Research, performance, and doing human geography: Some reflections on the diary-photograph, diary-interview method. *Environment and Planning A* 35 (November), 1993–2017.

Lau, C. and Li, Y.P. (2015) Producing a sense of meaningful place: Evidence from a cultural festival in Hong Kong. *Journal of Tourism and Cultural Change* 13 (1), 56–77.

Lau, C. and Li, Y.P. (2019) Analyzing the effects of an urban food festival: A place theory approach. *Annals of Tourism Research* 74, 43–55.

Laws, E. (1998) Conceptualising visitor satisfaction management in heritage settings: An exploratory blueprinting analysis of Leeds Castle, Kent. *Tourism Management* 19 (6), 545–554.

Lee, J.J., Kyle, G. and Scott, D. (2012) The mediating effect of place attachment on the relationship between festival satisfaction and loyalty to the festival hosting destination. *Journal of Travel Research* 51, 754–767.

Leed, E.J. (1991) *The Mind of the Traveller: From Gilgamesh to Global Tourism*. New York: Basic Books.

Leiper, N. (1990) *Tourism Systems: An Interdisciplinary Perspective*. Palmerston North: Massey University.

Lepp, A. and Harris, J. (2008) Tourism and national identity in Uganda. *The International Journal of Tourism Research* 10 (6), 525–536.

Lew, A.A. (2017) Tourism planning and place making: Place-making or placemaking? *Tourism Geographies* 19 (3), 448–466.
Li, C.C. (2004) Key speech in a report conference on national red tourism development planning Ootline (in Chinese, presented at the 'Report Conference on *National Red Tourism Development Planning Outline*'), Beijing, China.
Li, C.C. (2004) Fazhan shehuizhuyi xianjin wenhua [Developing progressive socialist culture]. *Yunnan Ribao*, 27 November [*Yunnan Daily*, 27 November]. See news.sohu .com/20041127/n223212125.shtml (accessed 29 November 2021).
Li, T.Y. and Wang, L.Y. (1988) *Luyou xue gailun* [*Of Tourism Studies*]. Tianjing, China: Nankai University Press.
Li, Y. (2011) Ethnic tourism and cultural representation. *Annals of Tourism Research* 38 (2), 561–585.
Li, Y.P. (2000) Geographical consciousness and tourism experience. *Annals of Tourism Research* 27 (4), 863–883.
Li, Y.P. (2002) The impact of tourism in China on local communities. *Asian Studies Review* 26 (4), 472–486.
Li, Y.P. (2003a) Heritage tourism: Contradictions between conservation and change. *Tourism and Hospitality Research* 4 (3), 247–261.
Li, Y.P. (2003b) Development of Nanshan Cultural Tourism Zone in Hainan, China: Achievements made and issues to be resolved. *Tourism Geographies* 5 (4), 436–445.
Li, Y.P. (2004) Exploring community tourism in China: The case of Nanshan Cultural Tourism Zone. *Journal of Sustainable Tourism* 12 (3), 175–193.
Li, Y. and Hinch, T. (1997) Ethnic tourism attractions and their prospect for sustainable development at two sites in China and Canada. *Asia-Pacific Journal of Tourism Research* 2, 5–18.
Li, Y.P. and Lo, L.B.R. (2005) Opportunities and constraints of heritage tourism in Hong Kong's changing cultural landscape. *Tourism and Hospitality Research* 5 (4), 322–345.
Li, Y.P. and Hu, Z.Y. (2008) Red tourism in China. *Journal of China Tourism Research* 4 (2), 156–171.
Li, Y.P. and Lau, C. (2022) Making or remaking people and places through festivals: An island tourism perspective. *International Journal of Event and Festival Management* 13 (3), 249–266.
Li, Y.P. and Mai, Z. (2022) Seeing China's social-cultural change through the eyes of tourists. *Journal of Tourism and Cultural Change* 31 (8), 44–57.
Li, Y.P., Lai, K. and Feng, X. (2007) Problem of 'guanxi' for actualizing community tourism: A case study of relationship networking in China. *Tourism Geographies* 9 (2), 115–138.
Li, Y.P., Hu, Z.Y. and Zhang, C.Z. (2010) Red tourism: Sustaining communist identity in a rapidly changing China. *Journal of Tourism and Cultural Change* 8 (1–2), 101–119.
Li, Y.P., Zhang, H., Zhang, D. and Abrahams, R. (2019) Mediating urban transition through rural tourism. *Annals of Tourism Research* 75 (2019), 152–164.
Li, Y.P., Lau, C. and Su, P. (2020) Heritage tourism stakeholder conflict: A case of a world heritage site in China. *Journal of Tourism and Cultural Change* 18 (3), 267–287.
Light, D. (2000a) Gazing on communism: Heritage tourism and post-communist identities in Germany, Hungary and Romania. *Tourism Geographies* 2 (2), 157–176.
Light, D. (2000b) An unwanted past: Contemporary tourism and the heritage of communism in Romania. *International Journal of Heritage Studies* 6 (2), 145–160.
Light, D. (2001) Facing the future: Tourism and identity-building in post-socialist Romania. *Political Geography* 20 (8), 1053–1074.
Lin, Y.H. and Lee, T.H. (2020) How the authentic experience of a traditional cultural festival affects the attendee's perception of festival identity and place identity. *International Journal of Event and Festival Management* 11 (3), 357–373.
Lincoln, Y.S. and Guba, E.G. (1985) *Naturalistic Inquiry*. Beverly Hills, CA: Sage.

Liu, M. (2001) Luyou di zhoubian xiangguan shequ de gongneng yu jiegou gengxin [Functional and structural renewal of rural communities surrounding tourism destinations]. *Huazhong shifan daxue xuebao* [*Academic Journal of Central China Normal University*] 35, 95.

Liu, W.H (2000) Guanyu shequ canyu luyou fazhan de yuogan sikao [Reflection on community participation in tourism development]. *Luyou Xuekan* [*Tourism Tribune*] 15, 41–47.

Liu, Z.L. (1992) Luyou dui jiedaidi de yingxiang [The impacts of tourism on a destination]. *Luyou Xuekan* [*Tourism Tribune*] 7, 52–55.

Logan, J.R. and Molotch, H. (2007) *Urban Fortunes: The Political Economy of Place*. Berkeley, CA: University of California Press.

Loi, K.I. and Pearce, P.L. (2012) Annoying tourist behaviors: Perspectives of hosts and tourists in Macao. *Journal of China Tourism Research* 8 (4), 395–416.

Lou, J.J. (2012) Chinatown in Washington, DC: The bilingual landscape. *World Englishes* 31 (1), 34–47.

Lowenthal, D. (1961) Geography, experience, and imagination: Towards a geographical epistemology. *Annals of the Association of American Geographers* 51 (3), 241–260.

Lowenthal, D. (1996) *Possessed by the Past: The Heritage Crusade and the Spoils of History*. New York: Free Press.

Lyotard, J.F. (1984) *The Post-Modern Condition: A Paper on Knowledge*. Manchester: Manchester University Press.

Ma, R. (1999) Population growth and urbanisation. In R.E. Gamer (ed.) *Understanding Contemporary China* (pp. 207–236). Boulder, CO: Lynne Rienner Publishers, Inc.

Ma, M. and Liu, Y. (2001) Tibet railway: A multi-angle analysis. *Chinese National Geography*, May (Special Issue), 22–27.

MacCannell, D. (1973) Staged authenticity: Arrangements of social space in tourist settings. *American Journal of Sociology* 79 (3), 589–603.

MacCannell, D. (1976) *The Tourist: A New Theory of the Leisure Class*. New York: Schocken Books.

MacCannell, D. (1989) *The Tourist: A New Theory of the Leisure Class*. New York: Schocken Books.

MacCannell, D. (1992) *Empty Meeting Ground: The Tourist Papers*. London: Routledge.

MacLeod, N.E. (2006) The placeless festival: Identity and place in the post-modern festival. In D. Picard and M. Robinson (eds) *Festivals, Tourism and Social Change: Remaking Worlds* (pp. 222–237). Clevedon: Channel View Publications.

Magliocco, S. (2001) Coordinates of power and performance: Festivals as sites of (re)presentation and reclamation in Sardinia. *Ethnologies* 23 (1), 167–188.

Mair, J., Chien, P.M., Kelly, S.J. and Derrington, S. (2021) Social impacts of mega-events: A systematic narrative review and research agenda. *Journal of Sustainable Tourism*. https://doi.org/10.1080/09669582.2020.1870989

Mandel, E. (1978) *Late Capitalism*. London: Verso.

Manen, M.V. (2014) *Phenomenology of Practice*. New York: Routledge.

Mannell, R. and Iso-Ahola, S. (1987) Psychological nature of leisure and tourism experience. *Annals of Tourism Research* 14 (3), 314–331.

Markusen, A. and Schrock, G. (2006) The distinctive city: Divergent patterns in growth, hierarchy and specialization. *Urban Studies* 43 (8), 1301–1323.

Marlow, C. (1998) *Research Methods for General Social Work*. Pacific Grove, CA: Brooks/Cole Publishing Company.

Mason, K.S. (2004) *Natural Museums: U.S. National Parks, 1872–1916*. East Lansing, MI: Michigan State University Press.

Mason, P. (2020) *Tourism Impacts, Planning and Management*. Abingdon: Routledge.

Massey, D.B. (1994) *Space, Place and Gender*. Cambridge: Polity Press.

Massey, D. (1995) The conceptualisation of place. In D.B. Massey and P. Jess (eds) *A Place in the World? Places, Cultures and Globalisation* (pp. 45–86). Oxford: Open University.

Massey, D. and Jess, P. (1995) *A Place in the World*. New York: Oxford University Press.

Mathieson, A. and Wall, G. (1982) *Tourism: Economic, Physical and Social Impacts*. London: Longman.

Mathieson, A. and Wall, G. (1997) *Tourism: Change, Impacts and Opportunities*. Harlow: Longman.

Mayo, E.J. (1975) Tourism and national parks: A psychographic and attitudinal study. *Journal of Travel Research* 14 (1), 14–18.

McArthur, S. and Hall, C.M. (1996) Interpretation: Principles and practice. In C.M. Hall and S. McArthur (eds) *Heritage Management in Australia and New Zealand: The Human Dimension* (pp. 88–106). Melbourne: Oxford University Press.

McLean, F. (1998) Museums and the construction of national identity: A review. *International Journal of Heritage Studies* 3 (4), 244–252.

Mercer, D.C. and Powell, J.M. (1972) *Phenomenology and Related Non-Positivistic Viewpoints*. Clayton: Monash University Publications in Geography.

Michaud, J. and Turner, S. (2006) Contending visions of a hill-station in Vietnam. *Annals of Tourism Research* 33 (3), 785–808.

Mill, R.C. and Morrison, A.M. (1992) *The Tourism System: An Introductory Text*. Englewood Cliffs, NJ: Prentice-Hall.

Mill, R.C. and Morrison, A.M. (2009) *The Tourism System*. Dubuque, IA: Kendall/Hunt Publishing.

Millard, R. (2022) WHO says China's zero-Covid strategy unsustainable. *Hong Kong Free Press*. See https://hongkongfp.com/2022/05/11/who-says-chinas (accessed 6 October 2022).

Ming Pao Daily (2000) *Ming Pao Daily News*, 20 October. See http://www.mingpaonews.com/200009/20/cfa2r.htm (accessed 20th October 2000).

Mitchell, R.E. (2001a) Community integration: Island tourism in Peru. *Annals of Tourism Research* 28 (1), 113–139.

Mitchell, T. (2001b) Making the nation: The politics of heritage in Egypt. In N. Alsayyad (ed.) *Consuming Tradition, Manufacturing Heritage* (pp. 212–239). London: Routledge.

Morgan, J. (1994) Wanuskewin: A meeting place for 6000 years. *Management Matters* 1994 (1), 8–11.

Morgan, N. and Pritchard, A. (1998) *Tourism Promotion and Power: Creating Images, Creating Identity*. Chichester: John Wiley and Sons.

Morishita, S.S. (2001) *Teodori: Cosmological Building and Social Consolidation*. Rome: Editrice Pontifica Universita Gregoriana.

Morrison, J. (1986) The wilderness solo: Solitude and recreation. Unpublished MA thesis, University of Alberta.

Moscardo, G. and Pearce, P.L. (1999) Understanding the tourists. *Annals of Tourism Research* 26 (2), 416–430.

Moser, A. and Korstjens, I. (2018) Series: Practical guidance to qualitative research. Part 3: Sampling, data collection and analysis. *European Journal of General Practice* 24 (1), 9–18.

Mostafanezhad, M. and Swain, M.B. (2019) Beyond anthropology: Ethnography in tourism studies. In N.M. Leite, Q.E. Castañeda and K.M. Adams (eds) *The Ethnography of Tourism: Edward Bruner and Beyond* (pp. 237–246). Lanham, MD: Lexington Books.

Mules, T. (1998) Events tourism and economic development in Australia. In D. Tyler, M. Robertson and Y. Guerrier (eds) *Managing Tourism in Cities: Policy, Process, and Practice* (pp. 195 – 214). Chichester: J. Wiley.

Munasinghe, H. (2005) The politics of the past: Constructing a national identity through heritage conservation. *International Journal of Heritage Studies* 11 (3), 251–260.

Murphy, P.E. (1985) *Tourism: A Community Approach*. New York: Methuen.

Nardi, P.M. (2014) *Doing Survey Research: A Guide to Quantitative Methods* (3rd edn). Boulder, CO: Paradigm Publishers.

Nash, D. (1996) *Anthropology of Tourism*. Oxford: Pergamon.

National Natural Science Foundation of China (2022) *China Natural Science Fund Project Guide 2020*. Beijing: National Natural Science Foundation of China: https://isisn.nsfc.gov.cn/egrantweb/

Ng, S.I., Chia, K.W., Ho, J.A. and Ramachandran, S. (2017) Seeking tourism sustainability: A case study of Tioman Island, Malaysia. *Tourism Management* 58 (1), 101–107.

Nicholls, S., Vogt, C. and Jun, S.H. (2004) Heeding the call for heritage tourism. *Parks & Recreation* 39 (9), 38–47.

Nickerson, N.P. (1996) *Foundations of Tourism*. Upper Saddle Hall, NJ: Prentice Hall Career and Technology.

Nielsen, N. and Wilson, E. (2012) From invisible to indigenous-driven: A critical typology of research in indigenous tourism. *Journal of Hospitality and Tourism Management* 19 (1), 67–75.

Nielsen-Pincus, M., Hall, T., Force, J.E. and Wulfhorst, J.D. (2010) Sociodemographic effects on place bonding. *Journal of Environmental Psychology* 30 (4), 443–454.

NRTCET (2008) *Zhongguo hongse lvyou fazhan baogao 2007* [*China Red Tourism Development Report 2009*]. Beijing: China Tourism Press.

NRTCET (2009) Red tourism. National Red Tourism Coordination Executive Team [NRTCET] (2008; 2009). www.redtourism.com.cn

NRTCET (2011) *Zhongguo hongse lvyou fazhan baogao 2009* [*China Red Tourism Development Report 2009*]. Beijing: China Tourism Press.

Nuryanti, W. (1996) Heritage and postmodern tourism. *Annals of Tourism Research* 23 (2), 249–260.

Oakes, T. (1995) Tourism in Guizhou: The legacy of internal colonialism. In A. Lew and L. Yu (eds) *Tourism in China: Geographic, Political, and Economic Perspectives* (pp. 203–222). Boulder, CO: Westview Press.

Oakes, T. (1998) *Tourism and Modernity in China*. London: Routledge.

Oakes, T. (2016) Villagizing the city: Turning rural ethnic heritage into urban modernity in southwest China. *International Journal of Heritage Studies* 22 (10), 751–765.

Oke, N., Sonn, C.C. and McConville, C. (2018) Making a place in footscray: Everyday multi-culturalism, ethnic hubs and segmented geography. *Identities* 25 (3), 320–338.

Oliveira, W., Baizerman, M. and Pellet, L. (1992) Street children in Brazil and their helpers: Comparative views on aspirations and the future. *International Social Work* 35 (2), 163–176.

Olsen, D. and Timothy, D.J. (2002) Contested religious heritage: Differing views of Mormon heritage. *Tourism Recreation Research* 27 (2), 7–15.

Ooi, C. (2005) The Orient responds: Tourism, Orientalism and the national museums of Singapore. *Review* 53 (4), 285–299.

O'Reilly, K. (2012) Ethnographic returning, qualitative longitudinal research and the reflexive analysis of social practice. *Sociological Review* 60 (3), 518–536.

Oriental News (2000) A38, 18 March 2000.

Osborne, J.W. (1990) Some basic existential-phenomenological research methodology for counsellors. *Canadian Journal of Counselling and Psychotherapy* 24 (2), 79–91.

Overton, J. and Murray, W.E. (2016) Fictive place. *Progress in Human Geography* 40 (6), 794–809.

Page, S. (1997) Urban tourism: Analyzing and evaluating the tourist experience. In C. Ryan (ed.) *The Tourist Experience: A New Introduction* (pp. 112–135). London: Cassell.

Paihangbang (2017) [Ranking list of population of Southeast Guizhou Prefecture in 2017]. *Chengshi > Renkou [City and Population]*. https://www.phb123.com/city/renkou/17110.html (accessed 8 October 2020).

Pain, R. (2004) Social geography: Participatory research. *Progress in Human Geography* 25 (5), 652–633.
Palmer, C. (1999) Tourism and the symbols of identity. *Tourism Management* 20 (3), 313–321.
Pan, S., Santos, C. and Kim, S. (2017) Promoting tourism, projecting power: The role of television commercials. *Journal of Travel & Tourism Marketing* 34 (2), 192–208.
Patton, M. (1982) Qualitative methods and approaches: What are they? *New Directions for Institutional Research* 34, 3–15.
Patton, M. (1990) *Qualitative Evaluation Methods*. Beverly Hills, CA: Sage.
Pearce, P. (1982) *The Social Psychology of Tourist Behaviour*. New York: Pergamon.
Peckham, R.S. (2003) Introduction: The politics of heritage and public culture. In R.S. Peckham (ed.) *Rethinking Heritage: Cultures and Politics in Europe* (pp. 1–16). London: I.B. Tauris.
Pengbai News (2020) Haidaoyou zaixian! Guoqing Zhuhai haidaoyou tupo 42,000 renci [Island heat reappears! Zhuhai Islands Tour exceeded 42,000 visitors during the National Day Holiday]. *Zgebgwu Zhuhai Fulian* [*Government Affairs. Zhuhai Women's Federation*], 8 October. See https://m.thepaper.cn/baijiahao_9487490 (accessed 11 November 2021).
People's Daily (2000) See http://english.peopledaily.com.cn/200005/23/eng20000523_41408.html (accessed 23 May 2000).
Peterson, Y.Y. (1995) The Chinese landscape as a tourist attraction: Image and reality. In A. Lew and L. Yu (eds) *Tourism in China: Geographic, Political, and Economic Perspectives* (pp. 141–154). Boulder, CO: Westview Press.
Picard, M. and Wood, R.E. (eds) (1997) *Tourism, Ethnicity, and the State in Asian and Pacific Societies*. Honolulu, HI: University of Hawaii Press.
Pickles, J. (1986) *Geography and Humanism*. Norwich: Geo Books.
Pierce, J., Martin, D.G. and Murphy, J.T. (2011) Relational place-making the networked politics of place. *Transactions of the Institute of British Geographers* 36 (1), 54–70.
Pizam, A. and Sussmann, S. (1995) Does nationality affect tourist behaviour? *Annals of Tourism Research* 22 (4), 901–917.
Poirier, R.A. (1995) Tourism and development in Tunisia. *Annals of Tourism Research* 22 (1), 157–171.
Polkinghorne, D.E. (1989) Phenomenological research methods. In R.S. Valle and S. Halling (eds) *Existential-Phenomenological Perspectives in Psychology: Exploring the Breadth of Human Experience* (pp. 41–62). New York: Plenum.
Pollio, H.R., Henley, T.B. and Thompson, C.J. (1997) *The Phenomenology of Everyday Life*. Cambridge: Cambridge University Press.
Poria, Y. and Ashworth, G. (2009) Heritage tourism: Current resource for conflict. *Annals of Tourism Research* 36 (3), 522–525.
Poria, Y., Butler, R. and Airey, D. (2001) Clarifying heritage tourism. *Annals of Tourism Research* 28 (4), 1047–1049.
Poria, Y., Biran, A. and Reichel, A. (2009) Visitors' preferences for interpretation at heritage sites. *Journal of Travel Research* 48 (1), 92–105.
Potter, T. (1993) A journey through weekend wilderness experiences. Unpublished doctorate dissertation, University of Alberta.
Pred, A. (1984) Place as historically contingent process: Structuration and the time-geography of becoming places. *Annals of the Association of American Geographers* 74 (2), 279–297.
Pretes, M. (2003) Tourism and nationalism. *Annals of Tourism Research* 30 (1), 125–142.
Prosser, R. (1994) Societal change and the growth in alternative tourism. In E. Cater and G. Lowman (eds) *Ecotourism: A Sustainable Option?* (pp. 19–28). New York: John Wiley and Sons Ltd.

Rahmat, N.H., Abdullah, N.A.T. and Kashiwazaki, K (2020) Linking situated learning in edutourism: A case study of Japanese undergraduates. *European Journal of Education Studies* 6 (12), 41–55.

Randstrom, R. and Deur, D. (1999) Reciprocal appropriation: Towards an ethics of cross-cultural research. In J.D. Proctor and D.M. Smith (eds) *Geography and Ethics: Journeys in a Moral Terrain* (pp. 237–250). London: Routledge.

Ratz, T. (2006) Interpretation in the House of Terror, Budapest. In M.K. Smith and M. Robinson (eds) *Cultural Tourism in a Changing World: Politics, Participation and (Re)presentation* (pp. 244–256). Clevedon: Channel View Publications.

Ray, M. (1994) The richness of phenomenology: Philosophic, theoretic, and methodological concerns. In M. Morse (ed.) *Critical Issues in Qualitative Research Methods* (pp. 117–135). Beverly Hills, CA: Sage.

Reisinger, Y. and Turner, L. (2003) *Cross-Cultural Behaviour in Tourism: Concepts and Analysis*. Oxford: Butterworth-Heinemann.

Relph, E.C. (1976) *Place and Placelessness*. London: Pion.

Relph, E.C. (1990) *The Phenomenological Foundations of Geography*. Discussion Paper Series. Department of Geography, University of Toronto.

Relph, E.C. (1997) Sense of place. In S. Hanson (ed.) *Ten Geographic Ideas that Changed the World* (pp. 205–226). New Brunswick, NJ: Rutgers University Press.

Relph, E.C. (2001) Place in geography. In N.J. Smelser and P.B. Baltes (eds) *International Encyclopedia of the Social and Behavioural Sciences* (pp. 11448–11451). Amsterdam: Elsevier.

Relph, E.C. (2016) The paradox of place and the evolution of placelessness. In R. Freestone and E. Liu (eds) *Place and Placelessness Revisited* (pp. 20–34). Abingdon: Routledge.

Ren, Q. (2010) Educational inequality under China's rural – urban divide: The *hukou* system and return to education. *Environment and Planning A* 42, 592–610.

Ren, Q. (2018) Educational inequality under China's rural–urban divide: The *hukou* system and return to education. *Environment and Planning A* 42, 592–610.

Richards, G. and Colombo, A. (2017) Rethinking the eventful city: Introduction. *Event Management* 21 (5), 527–531.

Richards, G. and Munsters, W. (2010) *Cultural Tourism Research Methods*. Wallingford: CABI.

Riley, R. (1993) Prestige worthy leisure travel behaviour. Unpublished doctorate dissertation, Texas A&M University.

Ritchie, J.R.B. (1984) Assessing the impact of hallmark events: Conceptual and research issues. *Journal of Travel Research* 23 (1), 2–11.

Roche, M. (2000) *Mega-events and Modernity: Olympics and Expos in the Growth of Global Culture*. London: Routledge.

Roche, M. (2003) Mega-events, time and modernity: On time structures in global society. *Time & Society* 12 (1), 99–126.

Rofel, L. (1999) *Other Modernities: Gendered Yearnings in China after Socialism*. Berkeley, CA: University of California Press.

Rogers, P. and Anastasiadou, C. (2011) Community involvement in festivals: Exploring ways of increasing local participation. *Event Management* 15 (4), 387–399.

Rohlen, T.P. (2000) *Hong Kong and the Pearl River Delta: 'One Country, Two Systems' In the Emerging Metropolitan Context*. Stanford, CA: Institute for International Studies, Asia/Pacific Research Centre, Stanford University.

Rose, G. (1995) Place identity: A sense of place. In D. Massey and P. Jess (eds) *A Place in the World? Places, Cultures and Globalisation* (pp. 88–132). New York: Oxford University Press.

Rosie, M., Alison, R. and Nigel, K. (2015) Colaizzi's descriptive phenomenological method. *The Psychologist* 28 (8), 643–644.

Ruzaitė, J. (2017) The linguistic landscape of tourism: Multilingual signs in Lithuanian and Polish resorts. *Eesti ja Soome-Ugri Keeleteaduse Ajakiri [Journal of Estonian and Finno-Ugric Linguistics]* 8 (1), 197–220.

Ryan, C. (1991) *Recreational Tourism: A Social Science Perspective*. London: Routledge.

Ryan, C. (1997) The chase of a dream, the end of a play. In C. Ryan (ed.) *The Tourist Experience: A New Introduction* (pp. 1–24). London: Cassell.

Ryan, C. and Montgomery, D. (1994) The attitudes of Bakewell residents to tourism and issues in community responsive tourism. *Tourism Management* 15 (5), 358–369.

Saar, M. and Palang, H. (2009) The dimensions of place meanings. *Living Reviews in Landscape Research* 3 (3), 1–24.

Said, E. (1978) *Orientalism, the Western Conception of the Orient*. New York: Vintage Books.

Sampson, K.A. and Goodrich, C.G. (2009) Making place: Identity construction and community formation through 'sense of place' in Westland, New Zealand. *Society and Natural Resources* 22 (10), 901–915.

Sauer, C.D. (1925) *The Morphology of Landscape*. Berkeley, CA: University of California Press.

Scannell, L. and Gifford, R. (2017) Place attachment enhances psychological need satisfaction. *Environment and Behaviour* 49 (4), 359–389.

Scheyvens, R. (2003) Local involvement in managing tourism. In S. Singh, D.J. Timothy and R.K. Dowling (eds) *Tourism in Destination Communities* (pp. 229–252). Oxford: CAB International.

Schiller, N.G. (1994) Introducing identities: Global studies in culture and power. *Identities* 1 (1), 1–6.

Schneekloth, L.H. and Shibley, R.G. (1995) *Placemaking: The Art and Practice of Building Communities*. New York: Wiley.

School of Geography and Planning (2004) Strategies for Huangshan scenic area tourism development (unpublished planning report). Sun Yat-sen University, Guangzhou.

Schroder-Esch, S., Kolbmuller, B. and Hassenflug, D. (2006) Introduction: Finishing a project, continuing a debate. In D. Hassenpflug, B. Kolbmuller and S. Schroder-Esch (eds) *Heritage and Media in Europe: Contributing towards Integration and Regional Development* (pp. 7–14). Weimar: Bauhaus-Universitat Weimar.

Seamon, D. (1979) *A Geography of the Life World: Movement, Rest and Encounter*. London: Grown Helm.

Seamon, D. (2000) *Dwelling, Place and Environment: Towards a Phenomenology of Person and World*. Malabar, FL: Krieger Publishing Company.

Seamon, D. (2013) Place attachment and phenomenology: The synergistic dynamism of place. In L.C. Manzo and P. Devine-Wright (eds) *Place Attachment: Advances in Theory, Methods, and Applications* (pp. 11–22). New York: Routledge.

Selwyn, T. (1996) Introduction. In T. Selwyn (ed.) *The Tourist Image* (pp. 1–32). Chichester: Wiley.

Shabanirad, E. and Marandi, S.M. (2015) Edward Said's Orientalism and the representation of Oriental women in George Orwell's 'Burmese Day'. *International Letters of Social and Humanistic Sciences Online* 60 (1), 22–33. doi:10.18052/www.scipress.com/ILSHS.60.22.

Shackel, P.A. (2001) Public memory and the search for power in American historical archaeology. *American Anthropologist* 103 (3), 655–670.

Shackel, P.A. (2005) Local identity, national memory, and heritage tourism: Creating a sense of place with archaeology. *Illinois Antiquity* 40 (3), 24–28.

Shaffer, M.S. (2001) *See America First: Tourism and National Identity 1880–1940*. Washington, DC: Smithsonian Institution Press.

Shao, X.Y. and Wang, X.D. (2006) Lun hongse lvyou mudidi de kaifa yu jianshe: Yi Shanxi Zuoquan County Matian kangzhan genjudi weili [The development of red

tourism destination: A case study of anti-Japanese War base in Matian County, Shanxi Province]. *Renwen Dili* [*Human Geography*] 21 (2), 32–36.
Sharpley, R. (2003) Tourism, modernisation and development on the island of Cyprus: Challenges and policy responses. *Journal of Sustainable Tourism* 11 (2–3), 246–265.
Sharpley, R. (2018) *Tourism, Tourists and Society*. Abingdon: Routledge.
Sharpley, R. (2021) Tourism and development theory: Which way now? *Tourism Planning & Development* 19 (1), 1–12.
Sharpley, R. and Ussi, M. (2014) Tourism and governance in small island developing states (SIDS): The case of Zanzibar. *International Journal of Tourism Research* 16 (1), 87–96.
Shenzhen Statistics Bureau (1999) *Basic Facts of Shenzhen's Tourism*. Shenzhen statistics. Shenzhen: Statistics Press. Shenzhen Trade Development Bureau. 2000. http://www.shenzhenwindow.net/business/rs-facts.htm
Shenzhen Trade Development Bureau (2000) *Basic Facts of Shenzhen's Tourism*. http://www.shenzhenwindow.net/business/rs-facts.htm
Shields, R. (1991) *Places on the Margin: Alternative Geographies of Modernity*. London: Routledge.
Shields, R. (1992) The individual, consumption cultures and the fate of community. In R. Shields (ed.) *Lifestyle Shopping: The Subject of Consumption* (pp. 99–113). London: Routledge.
Shim, C. and Santos, C.A. (2014) Tourism, place and placelessness in the phenomenological experience of shopping malls in Seoul. *Tourism Management* 45 (0), 106–114.
Shim, C., Santos, C.A. and Choi, M.J. (2013) Malling as a leisure activity in South Korea. *Journal of Leisure Research* 45 (3), 367–392.
Shim, C., Vargas, P.T. and Santos, C.A. (2015) Oriental imagery and American attitudes toward Asia: An exploratory tourism study. *Journal of Tourism and Cultural Change* 13 (2), 165–181.
Silvert, W. (2001) Modeling as a discipline. *International Journal of General Systems* 30 (3), 261–282.
Smith, A.D. (2001a) *Nationalism: Theory, Ideology, History*. Cambridge: Polity Press.
Smith, A. (2001b) Sporting a new image? Sport-based regeneration strategies as a means of enhancing the image of the city tourist destination. In C. Gratton and I.P. Henry (eds) *Sport in the City: The Role of Sport in Economic and Social Regeneration* (pp. 127–148). London: Routledge.
Smith, M. (1979) Geography, science and post-positivist modes of explanation. *Progress in Human Geography* 3 (3), 356–383.
Smith, R.M. (1982) *Learning How to Learn*. Chicago, IL: Follett.
Smith, S.L.J. (1995) *Tourism Analysis: A Handbook*. Harlow: Longman.
Smith, V.L. (1989) Eskimo tourism: Micromodels and marginal men. In V. Smith (ed.) *Hosts and Guests: Anthropology of Tourism* (pp. 55–82). Philadelphia, PA: University of Philadelphia Press.
Soja, E.W. (2000) *Postmetropolis: Critical Studies of Cities and Regions*. Oxford: Blackwell Publishers.
Sousa, D. (2014) Validation in qualitative research: General aspects and specificities of the descriptive phenomenological method. *Qualitative Research in Psychology* 11 (2), 211–227.
Spiegelberg, H. (1965) *The Phenomenological Movement: A Historical Introduction*. The Hague: Nijhoff.
Squire, S.J. (1994a) Gender and tourist experiences: Assessing women's shared meanings for Beatrix Potter. *Leisure Studies* 13 (3), 195–209.
Squire, S.J. (1994b) Accounting for cultural meanings: The interface between geography and tourism studies re-examined. *Progress in Human Geography* 18 (1), 1–16.
Staiff, R. (2014) *Re-imagining Heritage Interpretation*. Abingdon: Routledge.

Stapley, E., O'Keeffe, S. and Midgley, N. (2022) Developing typologies in qualitative research: The use of ideal-type analysis. *International Journal of Qualitative Methods* 21, 1–9.

Stark, F. (1953) *The Coast of Incense: Autobiography 1933–1939*. London: Murray.

Stern, E. and Krakover, S. (1993) The formation of a composite urban image. *Geographical Analysis* 25 (2), 130–146.

Stevenson, N. (2019) The street party: Pleasurable community practices and placemaking. *International Journal of Event and Festival Management* 10 (3), 304–318.

Stewart, D. (1990) *Exploring Phenomenology: A Guide to the Field and Literature*. Athens, OH: Ohio University Press.

Stillerman, J. and Salcedo, R. (2012) Transposing the urban to the mall: Routes, relationships, and resistance in two Santiago, Chile, shopping centers. *Journal of Contemporary Ethnography* 41 (3), 309–336.

Stone, L. (1989) Cultural cross-roads of community participation in development: A case from Nepal. *Human Organization* 48 (3), 206–213.

Stones, R. (2005) *Structuration Theory*. Basingstoke: Palgrave Macmillan.

Stratford, E., Baldacchino, G., McMahon, E., Farbotko, C. and Harwood, A. (2011) Envisioning the archipelago. *Island Studies Journal* 6 (2), 113–130.

Strauss, A.L. (1987) *Qualitative Analysis for Social Scientists*. Cambridge: Cambridge University Press.

Strauss, A.L. and Corbin, J. (1990) *Basics of Qualitative Research: Grounded Theory Procedures and Techniques*. Beverly Hills, CA: Sage.

Stritch, D. (2006) Archaeological tourism: A signpost to national identity. In I. Russell (ed.) *Images, Representations and Heritage: Moving beyond Modern Approaches to Archaeology* (pp. 43–60). New York: Springer.

Stronza, A. and Hunt, C. (2012) Vision of tourism: From modernisation to sustainability. *Practicing Anthropology* 34 (3), 19–22.

Strzelecka, M., Boley, B.B. and Woosnam, K.M. (2017) Place attachment and empowerment: Do residents need to be attached to be empowered? *Annals of Tourism Research* 66, 61–73.

Stutz, C. (2006) Mobility. In B. Warf (ed.) *Encyclopedia of Human Geography* (pp. 303–304). Thousand Oaks, CA: Sage Reference.

Su, M.M., Wall, G. and Wang, A. (2017) Yujiale fishing tourism and island development in Changshan Archipelago, Changdao, China. *Island Studies Journal* 12 (2), 127–142.

Su, X. (2018) Thousands flock to Cheung Chau Bun Festival as six-day heatwave in Hong Kong keeps others away from annual event. *South China Morning Post*. See https://www.scmp.com/news/hong-kong/community/article/2147212/thousands-flock-cheung-chau-bun-festival-six-day-heatwave (accessed 2 October 2021).

Su, X. and Teo, P. (2009) *The Politics of Heritage Tourism in China: A View from Lijiang*. Abingdon: Routledge.

Sullivan, C. and Forrester, M.A. (2019) *Doing Qualitative Research in Psychology: A Practical Guide*. Los Angeles, CA: Sage.

Sun, L.H. (1994) Deaths in China of 24 Taiwanese were no accident, Beijing says. *The Washington Post*, 19 April. See https://www.washingtonpost.com/archive/politics/1994/04/19/deaths-in-china-of-24-taiwanese-were-no-accident-beijing-says/0a7cac8c-12cd-41f5-8f4c-eace1878817d/ (accessed 16 November 2021).

Sutton, D. (2001) *Remembrance of Repasts: An Anthropology of Food and Memory*. Oxford: Berg Publishers.

Sutton, W. (1967) Travel and understanding: Notes on the social structure of touring. *International Journal of Comparative Sociology* 8 (2), 218–223.

Swain, M.B. (1989) Developing ethnic tourism in Yunnan, China: Shilin Sani. *Tourism Recreation Research* 14 (1), 33–39.

Swain, M.B. (1990) Commoditising ethnicity in Southwest China. *Cultural Survival Quarterly* 14 (1), 26–29.

Swain, M.B. (1993) Women producers of ethnic arts. *Annals of Tourism Research* 20 (1), 32–51.

Swain, M.B. (1995) A comparison of state and private artisan production for tourism in Yunnan. In A. Lew and L. Yu (eds) *Tourism in China: Geographic, Political, and Economic Perspectives* (pp. 223–233). Boulder, CO: Westview Press.

Tanasescu, A. (2006) Tourism, nationalism and post-communist Romania: The life and death of Dracula Park. *Journal of Tourism and Cultural Change* 4 (3), 159–175.

Tang, Y. and Liang, Y. (2022) Staged authenticity and nostalgia of mining tourists in the Jiayang mining Geo-park of China. *Journal of Tourism and Cultural Change*. https://doi.org/10.1080/14766825.2022.2090259.

Tasci, A.D., Gartner, W.C. and Cavusgil, S.T. (2007) Conceptualisation and operationalisation of destination image. *Journal of Hospitality & Tourism Research* 31 (2), 194–223.

Taşer, S. (2022) Social studies education's two main branches: The relationship between history and geography. *Journal for Educators, Teachers and Trainers* 13 (2), 159–173.

Thevenaz, P. (1962) *What is Phenomenology? and Other Essays*. Chicago, IL: Quadrangle Books.

Thorpe, J. (2012) *Temagami's Tangled Wild: Race, Gender, and the Making of Canadian Nature*. Vancouver: University of British Columbia Press.

Tian, K. (ed.) (2020) Juezhan juesheng tuoping gongjian – duzhan wei zhaimao pingkun xian: dadai miaozhai zoushang lvyou tuoping lu. [A decisive step – supervising the battle against poverty over the poverty stricken counties: Dadai Miao ethnic minority village on the way out of poverty through tourism]. *Gowest Network*. See http://news.cnwest.com/szyw/a/2020/10/04/19155467.html (accessed 7 October 2022).

Till, K.E. (2003) Construction sites and showcases: Mapping 'the new Berlin' through tourism practices. In S.P. Hanna and V.J. Del Casino (eds) *Mapping Tourism* (pp. 51–78). Minneapolis, MN: University of Minnesota Press.

Timothy, D.J. (1999) Participatory planning: A view of tourism in Indonesia. *Annals of Tourism Research* 26 (2), 371–391.

Timothy, D.J. (2005) *Shopping Tourism, Retailing and Leisure*. Clevedon: Channel View Publications.

Timothy, D.J. and Butler, R.W. (1995) Cross-border shopping: A North American perspective. *Annals of Tourism Research* 22 (1), 16–34.

Timur, S. and Getz, D. (2008) A network perspective on managing stakeholders for sustainable urban tourism. *International Journal of Contemporary Hospitality Management* 20 (4), 445–461.

Tisdell, C. and Wen, J. (1991) Foreign tourism as an element in PR China's economic development strategy. *Tourism Management* 12 (1), 55–67.

Toops, S. (1995) Tourism in Xinjiang: Practice and place. In A. Lew and L. Yu (eds) *Tourism in China: Geographic, Political, and Economic Perspectives* (pp. 179–202). Boulder, CO: Westview Press.

Tosun, C. and Jenkins, C.L. (1998) The evolution of tourism planning in Third World countries: A critique. *Progress in Tourism and Hospitality Research* 4 (2), 101–114.

Tosun, C., Timothy, D.J. and Öztürk, Y. (2003) Tourism growth, national development and regional inequality in Turkey. *Journal of Sustainable Tourism* 11 (2), 133–161.

Tresidder, R. (1999) Tourism in sacred places. In D. Crouch (ed.) *Leisure/Tourism Geographies: Practices and Geographical Knowledge* (pp. 137–148). London: Routledge.

Tribe, J. (2001) Research paradigms and the tourism curriculum. *Journal of Travel Research* 39 (4), 442–448.

Tsui, S. (2018) 140,000 visitors expected for HK Wine & Dine Festival, many from Mainland China. *South China Morning Post*. See https://www.scmp.com/news/hong-kong

/hong-kong-economy/article/2170241/140000-visitors-expected-biggest-hong-kong-wine-and (accessed 2 October 2021).
Tuan, T.H. and Navrud, S. (2008) Capturing the benefits of preserving cultural heritage. *Journal of Cultural Heritage* 9 (3), 326–337.
Tuan, Y.F. (1965) 'Environment' and 'world'. *The Professional Geographer* 17 (5), 6–8.
Tuan, Y.F. (1967) Attitudes toward environment: Themes and approaches. In D. Lowenthal (ed.) *Environmental Perception and Behavior* (pp. 4–17). Chicago, IL: University of Chicago.
Tuan, Y.F. (1975) Place: An experiential perspective. *Geographical Review* 65 (2), 151–164.
Tuan, Y.F. (1976) Humanistic geography. *Annals of the Association of American Geographers* 66 (2), 266–276.
Tuan, Y.F. (1977) *Space and Place: The Perspective of Experience.* Minneapolis, MN: University of Minnesota Press.
Tuan, Y.F. (1989) Surface phenomena and aesthetic experience. *Annals of the Association of American Geographers* 79 (2), 233–241.
Tuan, Y.F. (1993) *Passing Strange and Wonderful: Aesthetics, Nature, and Culture.* Washington, DC: Island Press.
Tuan, Y.F. (2001a) Life as a field trip. *Geographical Review* 91 (1), 41–45.
Tuan, Y.F. (2001b) *Space and Place: The Perspective of Experience.* Minneapolis, MN: University of Minnesota Press.
Tuan, Y.F. (2006) Humanistic geography: A personal view. *Progress in Human Geography* 25 (2), 1–7.
Tuan, Y.F. (2013) *Landscape of Fear.* Minnesota, MN: University of Minnesota Press.
Tunbridge, J.E. (1994) Whose heritage? Global problem, European nightmare. In G.J. Ashworth and P.J. Larkham (eds) *Building a New Heritage: Tourism, Culture, and Identity in the New Europe* (pp. 123–134). London: Routledge.
Tunbridge, J.E. (1998) The question of heritage in European cultural conflict. In B. Graham (ed.) *Modern Europe: Place, Culture and Identity* (pp. 236–258). London: Arnold.
Turner, B.S. (1993) Outline of a theory of human rights. *Sociology* 27 (3), 489–512.
Turner, L. and Ash, J. (1975) *The Golden Hordes: International Tourism and the Pleasure Periphery.* London: Constable.
Turok, I. (2009) The distinctive city: Pitfalls in the pursuit of differential advantage. *Environment and Planning A* 41 (1), 13–30.
Urry, J. (1990) *The Tourist Gaze: Leisure and Travel in Contemporary Society.* London: Sage.
Urry, J. (2006) Preface: Places and performances. In T.O. Claudio Minca (ed.) *Travels in Paradox: Remapping Tourism* (pp. vii–x). Lanham, MD: Rowman and Littlefield.
Urry, J. and Larsen, J. (2011) *The Tourist Gaze 2.0.* Los Angeles, CA: Sage Publications.
Usher, R., Bryant, I. and Johnston, R. (1997) *Adult Education and the Post-modern Change: Learning Beyond the Limits.* London: Routledge.
Valle, R.S. and Halling, S. (1989) *Existential-Phenomenological Perspectives in Psychology: Exploring the Breadth of Human Experience.* New York: Plenum.
Ven Passen, C. (1957) *The Classical Tradition of Modern Geography.* Groningen: J.B. Wolters.
Vidon, E.S. (2019) Why wilderness? Alienation, authenticity, and nature. *Tourist Studies* 19 (1), 3–22.
Waldinger, R. (2013) Immigrant transnationalism. *Current Sociology Review* 61 (5–6), 756–777.
Wang, N. (2000) *Tourism and Modernity: A Sociological Analysis.* Oxford: Pergamon.
Wang, H., Zhang, L., Cai, L. and Morrison, A. (1997) The domestic travel market in the Central Kingdom. In *New Frontiers in Tourism Research IX.* Conference Proceedings of the International Society of Travel and Tourism Educators, San Diego, CA.

Wang, X., Luo, R., Zhang, L. and Rozelle, S. (2017) The education gap of China's migrant children and rural counterparts. *The Journal of Development Studies* 53 (11), 1865–1881.

Wang, Y. (2008) Naturalising ethnicity, culturalising landscape: The politics of world heritage in China. Unpublished doctoral dissertation. Duke University.

Wang, Z.K. and Bennett, M. (2020) Anywhere but here: Experiences of islandness in Pearl River Delta island tourism. *Island Studies Journal* 15 (1), 205–222.

Watson, R.S. (1994) Memory, history, and opposition under state socialism: An introduction. In R.S. Watson (ed.) *Memory, History, and Opposition under State Socialism* (pp. 1–20). Sante Fe, NM: School of American Research Press.

Wayne, S. (1999) The tourism engine. *The China Business Review* (November–December), 38–41.

WCED (1987) *Report of the World Commission on Environment and Development: Our Common Future*. New York: WCED.

Weis, K., Chambers, C. and Holladay, P. (2021) Social-ecological resilience and community-based tourism in the commonwealth of Dominica. *Tourism Geographies* 23 (3), 1–21.

Weiler, B. and Hall, C.M. (eds) (1992) *Special Interest Tourism*. London: Belhaven Press.

Welch, C. and Piekkari, R. (2006) Crossing language boundaries: Qualitative interviewing in international business. *Management International Review* 46 (4), 417–437.

Whelan, C. (2016) Organisational culture and cultural change: A network perspective. *Journal of Criminology* 49 (4), 583–599.

WHP (1992a) Wanuskewin Heritage Park resource management plan (Wanuskewin unpublished raw data). WHP.

WHP (1992b) Wanuskewin Heritage Park training plan (Wanuskewin unpublished raw data). WHP.

Wight, C.A. and Lennon, J.J. (2007) Selective interpretation and eclectic human heritage in Lithuania. *Tourism Management* 28 (2), 519–529.

Wikimedia (2022) Wikimedia commons. See http://commons.wikimedia.org/wiki/File:Guangdong_subdivisions_-_Zhuhai.svg (accessed 1 May 2022).

Wikipedia (2020) *Guizhou*. Wikipedia, the free encyclopedia. See https://en.wikipedia.org/wiki/Qiandongnan_Miao_and_Dong_Autonomous_Prefecture (accessed 12 November 2020).

Williams, D.R. and Vaske, J.J. (2003) The measurement of place attachment: Validity and generalization of a psychometric approach. *Forest Science* 49 (6), 830–840.

Williams, S. (1998) *Tourism Geography*. London: Routledge.

Wilson, D. (1996) Paradoxes of tourism in Goa. *Annals of Tourism Research* 24 (1), 52–75.

Wise, N., Badurina, J.D. and Peric, M. (2021) Assessing residents' perceptions of urban placemaking prior to hosting a major cultural event. *International Journal of Event and Festival Management* 12 (1), 51–69.

Woelfel, J. (1998) CATPAC II (Versions 2.0): User's guide. https://academic.csuohio.edu/kneuendorf/c63309/CATPAC2Manual.pdf

Wood, R.E. (1984) Ethnic tourism, the state, and cultural change in Southeast Asia. *Annals of Tourism Research* 11 (3), 353–374.

Woodley, A. (1993) Tourism and sustainable development: The community perspective. In J.G. Nelson, R.W. Butler and G. Wall (eds) *Tourism and Sustainable Development: Monitoring, Planning, Managing* (pp. 135–147). Waterloo: University of Waterloo, Heritage Resource Centre.

Worden, N. (2001) 'Where it all began': The representation of Malaysian heritage in Melaka. *International Journal of Heritage Studies* 7 (3), 199–218.

World Travel and Tourism Council (1995) *Agenda 21 for the Travel and Tourism Industry: Towards Environmentally Sustainable Development*. London: World Travel and Tourism Council.

Wright, J.K. (1947) *Human Nature in Geography*. Cambridge, MA: Harvard University Press.

WTO (2012) *Tourism and Intangible Cultural Heritage*. Madrid: World Tourism Organisation.

WTTC (2021) Economic impact 2021. See https://wttc.org/Portals/0/Documents/Reports/2021/Global%20Economic%20Impact%20and%20Trends%202021.pdf?ver=2021-07-01-114957-177 (accessed 26 September 2021).

Wu, B.H. (2001) *Quyu luyou guihua yuanli* [*Principles of Regional Tourism Planning*]. Beijing: China Tourism Press.

Wu, R.W. (2000) Luyou guihua de fazhan licheng yu qushi [Evolution and trend of tourism planning]. *Nongcun shengtai huanjing* [*Ecological Environment of the Countryside*] 16, 38.

Xiao, X., Zhang, J. and Lu, J.Y. (2012a) Jumin dui Zunyi hongse lv you shangyejie Zhong shufa jingguan de rentongdu yanjiu [Research on residents' identity of calligraphy landscape in Zunyi Red Tourism Commercial Street]. *Dili Kexue Jinzhan* [*Progress in Geographical Science*] 31 (8), 1080–1086.

Xiao, X., Zhang, J. and Sun, Q. (2012b) Shufa jingguan lvyoudi youke difanggan yingxiangyinsu fenxi: Yi Shaanxi Hanzhong Shunen Shisanpin weili [Analyzing the influence factors of tourists' sense of place in calligraphy scenic spots: A case study of the Thirteen Products of Shimen in Hanzhong, Shaanxi, China]. *Ren Wen Di Li* [*Human Geography*] 27 (6), 130–136.

Xinhua News (2006) Xiang Gang shangchang 3%–15% keliu laizi neidi [3–15% of shoppers in malls in Hong Kong from Mainland China]. *Xinhua News*, 31 December. See http://big5.xinhuanet.com/gate/big5/news.xinhuanet.com/tai_gang_ao/2006-12/31/content_5554285.htm (accessed 8 January 2014).

Xiuning Tourism Bureau (2004) Basic information about Xiuning and Wan'an. Personal communications faxed between the researcher and Xiuning Tourism Bureau.

Xue, L. and Zhang, Y. (2020) The effect of distance on tourist behaviour: A study based on social media data. *Annals of Tourism Research* 82 (4), 1–12. https://doi.org/10.1016/j.annals.2020.102916

Yang, H., Liu, C., Liu, Y. and Duan, Y. (1999) *Man Chun Man Village at the Cross-Road: Conservation and Change of Ethnic Cultures in the Tourism Gold Rush*. Kunming: Yunnan University Press.

Yang, K.Z. (2003) 31 shengshi jinji shili duibi [Comparison and contrast of economic development potentials among China's 31 provinces and municipalities]. See http://finance.sina.com.cn/guest69.html

Yasothornsrikul, P. and Bowen, D. (2015) Tourism demonstration and value change. *International Journal of Tourism Research* 17 (1), 96–104.

Yiannakis, A. and Gibson, H. (1992) Roles tourists play. *Annals of Tourism Research* 19 (2), 287–303.

Younge, G. (2019) The politics of identity: From potential to pitfalls, and symbols to substance. *Identities* 26 (1), 1–11.

Yu, W.J., Zhang, Y.J., Zhou, W.Q., Wang, W.M. and Tang, R. (2019) Urban expansion in Shenzhen since 1970s: A retrospect of change from a village to a megacity from the space. *Physics and Chemistry of the Earth, Parts A/B/C* 110, 21–30.

Zaner, R.M. (1971) *The Problem of Embodiment: Some Contributions to a Phenomenology of the Body*. The Hague: Nijhoff.

Zhang, C.X., L'Espoir Decosta, P. and McKercher, B. (2015) Politics and tourism promotion: Hong Kong's myth making. *Annals of Tourism Research* 54, 156–171.

Zhang, C., Xiao, H.G., Morgan, N. and Ly, T.P. (2018) Politics of memories: Identity construction in museums. *Annals of Tourism Research* 73, 116–130.

Zhang, E.H. (2008) From communism to consumerism, from revolution to recreation: Red tourism in China. *China 21* 29, 163–165 (in Japanese).

Zhang, G. (1995) China's tourism since 1978: Policies, experiences, and lessons learned. In A. Lew and L. Yu (eds) *Tourism in China: Geographic, Political, and Economic Perspectives* (pp. 3–17). Boulder, CO: Westview Press.

Zhang, G.R. (1998) Guanyu luyouye de 21 shiji yicheng: Shixian yu huanjing xiang shiyin de kechixu fazhen [Agenda 21 for the travel and tourism industry: Towards environmentally sustainable development]. *Luyou Xuekan* [*Tourism Tribune*] 13 (2), 3–5.

Zhang, J.F. (2005) Fazhan hongse lvyou zhi wojian [Developing red tourism – my point of view]. *Hezuo jinji yu keji* [*Cooperative Economy and Science and Technology*] 9 (10), 41–42.

Zhang, P. (2019) Orientalism revisited in the Chinese context. *Linguistics and Literature Studies* 7 (6), 250–257.

Zhang, Y. (1957) *Zhongguo jindai nongyeshi ziliao* [*Source Materials on the Agricultural History of Modern China*]. Beijing: Sanlian Shudian (Joint Publications).

Zhang, Y. (1990) *Luyou jingji gongzuo shouce* [*Work Manual of Tourism Economy*]. Beijing: China Encyclopedia Publishing House.

Zhao, L.Y. (2021) China's 'economic miracle' and the universal modernization model. *Modern China* 48 (1), 53–72.

Zhao, Z.Y. (1987) Advance along the road of socialism with Chinese characteristics. In *Documents of the 13th National Congress of the Communist Party of China*. Beijing: Foreign Languages Press.

Zhao, Y. and Liu, B. (2020) The evolution and new trends of China's tourism industry. *National Academy Review* 2 (4), 337–353.

Zhiyan Wisdom Group (2021) 2021–2027 Zhongguo Zhuhai lvyouhangye shichangjingzheng fenxi ji qianjing fenxi baogao [Analysis of market competition and development strategy of tourism industry in Zhuhai, China from 2021 to 2027]. *Chanye Yanjiu* [*Industry Research*]. See www.chyxx.com (accessed 6 September 2021).

Zhu, S., Luo, Y., Aziz, N., Jamal, A. and Zhang, Q. (2022) Environmental impact of the tourism industry in China: Analyses based on multiple environmental factors using novel Quantile Autoregressive Distributed Lag model. *Economic Research-Ekonomska Istraživanja* 35 (1), 3663–3689.

Zhu, T. (2001) Nationalism and China's foreign policy. *The China Review* 1 (1), 1–28.

Zhuang, D.Q. (2006) Discussion on several problems regarding sustainable development of red tourism. *Journal of Jiangxi University of Finance and Economics* 47 (5), 95–98.

Zhuang, L., Taylor, T., Beirman, D. and Darcy, S. (2017) Socially sustainable ethnic tourism: A comparative study of two Hakka communities in China. *Tourism Recreation Research* 42 (4), 467–483.

Zhuhai City Government (2017) *Zhuhai huwai guanggao sheshi shezhi ni zhezhi qufen guanli* [Zhuhai outdoor advertising facilities are planned to be managed by districts]. See http://www.sohu.com/a/151066739_119778 (accessed 11 November 2011).

Index

Figures are shown in *italics*, tables in **bold**.

For Product Safety Concerns and Information please contact our EU Authorised Representative:

Easy Access System Europe

Mustamäe tee 50

10621 Tallinn

Estonia

gpsr.requests@easproject.com

www.ingramcontent.com/pod-product-compliance
Ingram Content Group UK Ltd.
Pitfield, Milton Keynes, MK11 3LW, UK
UKHW021836150926
8656IPUK00003B/14

* 9 7 8 1 8 4 5 4 1 8 9 0 8 *